Controversies in Acute Kidney Injury

Contributions to Nephrology

Vol. 174

Series Editor

Claudio Ronco Vicenza

Controversies in Acute Kidney Injury

Volume Editors

John A. Kellum Pittsburgh

Claudio Ronco Vicenza

Jean-Louis Vincent Brussels

31 figures, 3 in color, and 16 tables, 2011

KARGER Basel · Freiburg · Paris · London · New York · New Delhi · Bangkok · Beijing · Tokyo · Kuala Lumpur · Singapore · Sydney

Contributions to Nephrology
(Founded 1975 by Geoffrey M. Berlyne)

John A. Kellum
Department of Critical Care Medicine
University of Pittsburgh
Pittsburgh, PA 15261
USA

Claudio Ronco
Department of Nephrology, Dialysis & Transplantation
International Renal Research Institute Vicenza
S. Bortolo Hospital, Viale Rodolfi, 37
36100 Vicenza, Italy

Jean-Louis Vincent
Department of Intensive Care
Erasme Hospital
Université libre de Bruxelles
Route de Lennik 808
1070 Brussels
Belgium

Library of Congress Cataloging-in-Publication Data

Controversies in acute kidney injury / volume editors, John A. Kellum, Claudio Ronco, Jean-Louis Vincent.
p. ; cm. -- (Contributions to nephrology, ISSN 0302-5144 ; v. 174)
Includes bibliographical references and indexes.
ISBN 978-3-8055-9810-1 (hard cover : alk. paper) -- ISBN 978-3-8055-9811-8 (e-ISBN)
I. Kellum, John A. II. Ronco, C. (Claudio), 1951- III. Vincent, J. L. IV. Series: Contributions to nephrology ; v. 174. 0302-5144
[DNLM: 1. Acute Kidney Injury--physiopathology. 2. Acute Kidney Injury--therapy. 3. Renal Replacement Therapy. W1 CO778UN v.174 2011 / WJ 342]
LC-classification not assigned
617.4'61044--dc23

2011025944

Bibliographic Indices. This publication is listed in bibliographic services, including Current Contents® and Index Medicus.

Drug Dosage. The authors and the publisher have exerted every effort to ensure that drug selection and dosage set forth in this text are in accord with current recommendations and practice at the time of publication. However, in view of ongoing research, changes in government regulations, and the constant flow of information relating to drug therapy and drug reactions, the reader is urged to check the package insert for each drug for any change in indications and dosage and for added warnings and precautions. This is particularly important when the recommended agent is a new and/or infrequently employed drug.

www.karger.com
Printed in Switzerland on acid-free and non-aging paper (ISO 9706) by Reinhardt Druck, Basel
ISSN 0302–5144
ISBN 978–3–8055–9810–1
e-ISBN 978–3–8055–9811–8

Contents

Renal Support

Kellum JA, Ronco C, Vincent J-L (eds): Controversies in Acute Kidney Injury.
Contrib Nephrol. Basel, Karger, 2011, vol 174, pp 1–3

Unveiling Current Controversies in Acute Kidney Injury

John A. Kellum[a] · Claudio Ronco[b] · Jean-Louis Vincent[c]

[a]Department of Critical Care Medicine, University of Pittsburgh, Pittsburgh, USA, [b]Department of Nephrology, Dialysis & Transplantation, International Renal Research Institute Vicenza, S. Bortolo Hospital, Vicenza, Italy, [c]Department of Intensive Care, Erasme Hospital, Université libre de Bruxelles, Brussels, Belgium

Acute kidney injury (AKI) is a deadly and incompletely understood disorder in which sudden impairment of kidney function occurs secondary to one or more of a variety of underlying conditions. Mortality associated with AKI is very high and treatment is unsatisfactory. The condition primarily affects acutely ill and injured patients and disproportionably affects the elderly. Many of those that survive have permanent kidney failure and other long-term morbidities, which may include cardiovascular disease and immune dysfunction. While the term 'acute kidney injury' dates back to the early 20th century, when it was used in reference to acute mercury poisoning, it has only recently been applied to describe impaired kidney function in the setting of critical illness. In 2002, during an Acute Dialysis Quality Initiative (ADQI) conference held in Vicenza, AKI was defined using the now widely accepted consensus criteria known as RIFLE (Risk, Injury, Failure, Loss, End-Stage Kidney Disease). AKI replaced the term acute renal failure in part because of the recognition that acute impairment in renal function, even when relatively mild and far less than frank failure, was associated with worse clinical outcomes. Criteria for AKI were therefore set at small changes in serum creatinine or urine output. Thanks to consensus criteria for AKI, we now know that this condition is very common, occurring in as many as two thirds of ICU patients and about 2,100 per million population, and is associated with dramatic reductions in survival – as much as a 3- to 8-fold decrement at hospital discharge compared to controls without AKI.

With the increasing body of information about AKI, it becomes more and more evident that we need to find an answer to some crucial questions, e.g. 'What are exactly the so-called pre-renal syndromes?', 'Do we know in detail the pathogenesis of AKI?', 'Are critically ill patients dying *of* or *with* AKI?', and 'What is the most appropriate management of AKI?'.

It seems that classic distinctions such as 'prerenal and 'renal' AKI may be questionable today or at least they should be scrutinized with a critical approach.

For example, if the prerenal state is very transient and/or mild, and when it occurs in the setting of normal baseline renal function, it may appear to be well tolerated. However, emerging evidence suggests that the prerenal state is precarious. First, it may potentiate renal toxicity from radiocontrast or other nephrotoxins. Second, renal impairment may lead to volume overload, acid-base and electrolyte imbalance, immune dysfunction, coagulation abnormalities, abnormal drug elimination, and direct effects on the function of various organs. Indeed renal impairment results in multiple organ failure. Finally, when severe or prolonged, or perhaps even when mild and transient but in an already compromised kidney, the prerenal state can lead to direct kidney damage.

Several other areas of controversy exist in the field of AKI. One important area concerns our understanding of the pathogenesis of AKI. Epidemiologic evidence suggests that AKI is not a single disease, but a syndrome comprised of multiple, often coexisting, etiologies. The most common forms of AKI appear to be nonischemic and arise in settings such as sepsis and heart failure. Early AKI may be purely functional and reversible, but it soon gives way to tissue injury and a complex array of vascular, metabolic and inflammatory changes. Furthermore, the kidney may be an innocent bystander injured by the very toxins that it filters from the blood in the setting of remote tissue injury and infection. Cytokines, free radicals and other damage-associated molecular patterns may initiate AKI. These same triggers may also lead to so-called 'maladaptive repair' mechanisms that can cause further damage particularly in the most susceptible patients such as the elderly and those with chronic kidney disease. On the other hand, the concept of acute tubular necrosis as the main pathogenetic picture in AKI seems to be questioned by the common findings that renal blood flow may not be reduced, at least in sepsis, and other types of cell damage may occur. The concomitant presence of inflammation seems to be a requirement to initiate and extend tissue damage and to activate multiple organ crosstalk responsible for the high morbidity and mortality associated with AKI.

Studies conducted in animals and humans have displayed their limits to elucidate all complex mechanisms involved in this disorder. Therapies that work in specific experimental models may have little or no efficacy when translated into the clinical realm where overlapping etiologies is the rule. There may be no simple solution for the complex problem that is AKI. Instead it may be possible to develop a suite of therapies to attenuate the many different inciting factors and to produce effective countermeasures that facilitate resolution of injury and promote recovery of function. Novel biomarkers for early detection of AKI and for predicting the course of disease in humans are being developed and, thus, we will soon have better ways to apply the right therapies to the right patients.

An interesting distinction is made today between renal replacement and renal support. Hemodialysis and hemofiltration are being delivered earlier and to more severely ill patients than ever before. In several cases this is done with the intent to support specific clinical conditions rather than to replace lost function of the kidney. There is some evidence that survival for patients with AKI is improving, though it is still quite poor in the most critically ill. Renal support may well be the bridge to recovery, but innovation is lagging – we have seen little change in the way we provide support over the last 3 decades. If we are to expect better outcomes, we will need to develop better therapies.

Increased hemofiltration volumes and new technologies such as higher cut-off membranes, plasma filtration and adsorption, and new sorbent devices are potential solutions to improve renal support. Technologies for extracorporeal removal of larger microbial toxins such as endotoxin are also becoming available. Finally, it must be recognized that AKI is usually part of multiorgan failure syndrome and extracorporeal support may also target fluid overload and heart failure, extracorporeal CO_2 removal for combined kidney and lung support, albumin dialysis for liver support, and other techniques unified under the umbrella of MOST (multiple organ support therapy). Such therapies aim to improve organ function and decrease the severity of organ damage.

In spite of significant advances, new challenges are also appearing in the clinical arena. We need to better understand why AKI occurs and to develop new methods to treat it. We need to improve outcomes and find ways to mitigate organ damage and stimulate organ recovery. The interesting and growing cooperation among specialists of different disciplines may well be the key to moving to a more holistic approach for the patient suffering from AKI. The process has begun and it is our responsibility to maintain it over the years to come.

Claudio Ronco
Department of Nephrology, Dialysis & Transplantation
International Renal Research Institute Vicenza
S. Bortolo Hospital, Viale Rodolfi, 37, 36100 Vicenza, Italy
Tel. +39 444 75 38 69, E-mail: cronco@goldnet.it

Kellum JA, Ronco C, Vincent J-L (eds): Controversies in Acute Kidney Injury.
Contrib Nephrol. Basel, Karger, 2011, vol 174, pp 4–11

Analysis of the Prerenal Contributions to Acute Kidney Injury

Roland C. Blantz · Prabhleen Singh

University of California San Diego, Division of Nephrology-Hypertension, Veterans Affairs San Diego Healthcare System, and O'Brien Center for Acute Kidney Injury Research, San Diego, Calif., USA

Abstract

Acute kidney injury (AKI) occurs frequently in hospitalized patients, and prerenal mechanisms contribute significantly to the pathogenesis of AKI. Prerenal contributions to renal dysfunction may be transient and reversible, as in volume depletion, or more persistent as observed with heart failure and liver disease. They can also act as a precursor to parenchymal kidney damage. The reductions in glomerular filtration rate are largely shared by all nephrons and are primarily mediated by reductions in nephron plasma flow and decreases in the glomerular ultrafiltration coefficient. Studies in animals suggest that adrenergic activity and angiotensin II (Ang II) are the dominant hormonal influences that independently and synergistically impact the determinants of glomerular filtration. Interactions between individual adrenoreceptors and Ang II are complex and significant. Tubular injury can also activate prerenal mechanisms via the tubuloglomerular feedback system. The effects of adrenergic and Ang II activities are counteracted by actions of nitric oxide and prostaglandins within the kidney. Bidirectional regulatory influences occur between the vasoconstrictor and vasodilatory hormonal systems. Understanding these prerenal mechanisms and the role of endogenous and exogenous vasoconstrictor and vasodilator hormones is important in the prevention, therapy and recovery of AKI in critically ill patients who commonly encounter it.

Acute kidney injury (AKI) remains a common event in critically ill hospitalized patients. Although 'acute tubular necrosis' or parenchymal kidney injury is thought to be the major cause of AKI in these patients, prerenal mechanisms contribute substantially to the pathogenesis of AKI [1, 2]. Prerenal mechanisms can contribute to AKI in various patterns. The reduction in glomerular filtration rate (GFR) may be transient albeit severe, as observed in massive volume depletion, but can be rapidly corrected by appropriate therapy. It may be more persistent in diseases of organs distant from the kidney such as congestive heart failure [3] and end-stage liver failure. Finally, it can present as a prodrome in which a prerenal condition quickly develops into parenchymal AKI, such as in sepsis and postsurgical conditions (e.g. following cardiac surgery) [2]. In this summary of prerenal contributions to AKI, we will describe (1) the determinants of glomerular ultrafiltration which commonly contribute to the reduction in GFR, (2) provide evidence for the contributions of endogenous neurohumoral substances that affect GFR and (3) describe the interactions among the multiple endogenous factors which act in the aggregate to generate the prerenal state [4].

Determinants of Ultrafiltration

Generally, in prerenal AKI there is a homogenous reduction in single-nephron GFR (SNGFR) in nearly all nephrons of the kidney. This is in contradistinction to parenchymal AKI or 'acute tubular necrosis' in which the effects on nephrons may be heterogenous and in which primary reductions in glomerular ultrafiltration may account for a modest component (approx. 50%) of the reduction in kidney GFR, the remainder relating to tubular backleak of solute and tubular obstruction [5]. Therefore, the question arises, 'By what mechanisms is the GFR per nephron reduced in prerenal AKI?' Since we are dealing with factors that impact primary glomerular ultrafiltration (in contradistinction to structurally based mechanisms such as tubular obstruction and backleak of solutes), there are four factors to consider: (1) the rate of nephron plasma flow; (2) the glomerular capillary hydraulic pressure gradient, ΔP; (3) the glomerular ultrafiltration coefficient, LpA or K_f; and (4) systemic oncotic pressure [4, 6]. The last factor is rarely a consideration in prerenal AKI [7].

Profound volume depletion, which is commonly examined and easy to recognize, is one of the clinical conditions that generates prerenal AKI [8–10]. More than 30 years ago, we examined the glomerular hemodynamic mechanisms that decrease whole kidney and SNGFR in an experimental model of volume or NaCl depletion, produced by administration of a loop diuretic, furosemide, on zero NaCl dietary intake [8]. We utilized Munich-Wistar rats possessing glomeruli on the surface of the kidney, permitting direct measurement of glomerular capillary hydraulic pressure and measurement of filtration fraction and nephron plasma flow in surface nephrons. In these rats, blood pressure was relatively well maintained; therefore, the animals were not hypotensive, only volume depleted. SNGFR was substantially lowered because of a major reduction in both the nephron plasma flow and the glomerular ultrafiltration coefficient, LpA. Both afferent (AR) and efferent (ER) arteriolar resistances increased during volume depletion. The ΔP was increased in part due to the greater incremental change in ER in comparison to AR and in part due to a major reduction in LpA, a major conductance pathway for water and small solutes. This increase in ΔP limits the fall in SNGFR.

The pathophysiology of prerenal conditions such as those associated with congestive heart failure or liver cirrhosis is also dominated by the changes in nephron plasma flow due to effects of endogenous neural and hormonal factors on the major resistors (afferent and efferent arterioles) in the kidney. In this manner, the determinants of glomerular ultrafiltration in these conditions are similar to those observed in the salt depletion described above. There is also a prerenal condition associated with calcineurin inhibitors such as cyclosporine (an immunosuppressive commonly used in organ transplantation) that differs to some extent in terms of glomerular hemodynamics. These agents produce a prerenal form of AKI, rapidly reversible by lowering the dose and replacing volume deficits. This condition is unique in that a reduction in ΔP is the dominant hemodynamic cause for a reduction in SNGFR [11]. This occurs as a consequence of modest afferent vasoconstriction, but with a fixed efferent arteriolar resistance, resulting in major reductions in glomerular capillary pressure and ΔP. The events appear to correlate with low renal angiotensin (Ang) II levels, resulting in reduced capacity to constrict the efferent arteriole [12].

Neurohormonal Interactions and Their Contribution to Prerenal Acute Kidney Injury

A variety of endogenous neurohumoral substances participate in the pathogenesis of prerenal azotemia, and usually arise in response to reduced effective circulating volume. As observed in figure 1, substances such as Ang II as part of the renin-angiotensin-aldosterone axis, renal adrenergic nerves, and antidiuretic hormone can all participate, primarily via influences on renal plasma flow and reductions in the glomerular ultrafiltration coefficient [2]. These actions are opposed by vasodilators, and experimental evidence supports the predominant role of nitric oxide (NO) and the prostaglandins as the counteractive hormones in the regulation of renal hemodynamics.

In prerenal azotemia due to salt depletion or congestive heart failure, increased intrarenal Ang II activity is observed. Ang II increases both the afferent and efferent vascular resistances in the nephron; however, the effects on the ER predominate early on, which increases ΔP and maintains GFR [13]. At later stages, both ER and AR undergo severe vasoconstriction leading to decreased nephron plasma flow. In addition, LpA is also lowered, which also contributes to the reduction in GFR. In animal models of salt depletion, administration of angiotensin-converting enzyme (ACE) inhibitors or angiotensin receptor blockers restores normal values for nephron plasma flow, LpA and ΔP [8, 10]. These effects are similar to the status that occurs after volume status is acutely restored

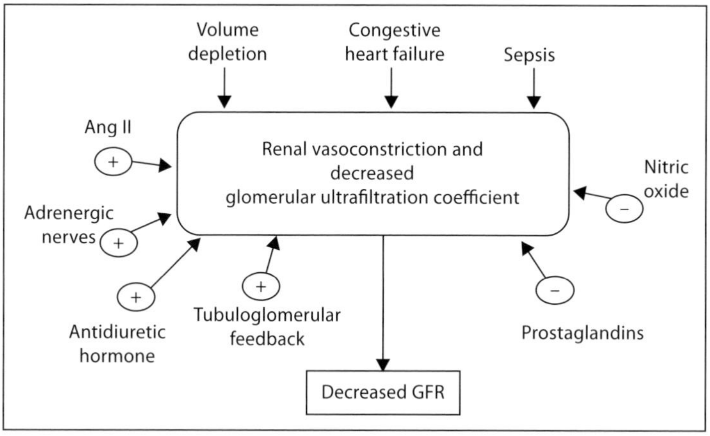

Fig. 1. Hormonal control of renal vascular resistance. Prerenal mechanisms that contribute to decreased GFR. This condition commonly derives from three major clinical conditions: volume depletion, congestive heart failure and sepsis. These conditions of functional kidney failure or AKI lead to renal vasoconstriction and reduced nephron plasma flow and decreased glomerular ultrafiltration coefficient, producing reductions in GFR without damage to the kidney parenchyma. Hormonal influences dictating these events include angiotensin II, adrenergic nerves and antidiuretic hormone, as well as tubuloglomerular feedback activity under certain circumstances. These largely vasoconstrictor systems are intimately interrelated and are counteracted effectively in the kidney by vasodilatory systems such as local nitric oxide and prostaglandins.

by plasma volume expansion. In fact, acute restoration of volume status greatly reduced plasma renin activity [9].

Given the findings above, one would be tempted to conclude that Ang II generation is the sole cause for the reduction in GFR during volume depletion. However, renal adrenergic nerve activity is also greatly increased during salt depletion. We therefore examined the relative contributions of renal adrenergic activity and Ang II to glomerular hemodynamics in salt depletion, with renal denervation and administration of ACE inhibitors, respectively [10]. Both renal denervation and ACE inhibition restored values for afferent arteriolar resistance to normal and increased nephron plasma flow. However, as renal denervation exerted a stronger effect on the efferent arteriole than ACE inhibition, this maneuver was more effective in restoring the normal SNGFR.

To further clarify the interactions between Ang II and adrenergic nerve activity, we examined the effects of direct renal nerve stimulation (RNS) on glomerular hemodynamics in the euvolemic rat [14]. Nerves coursing along the renal artery were gently dissected free and stimulated with an electrode during renal micropuncture to assess the glomerular hemodynamics. SNGFR and nephron plasma flow was decreased by approximately 25–30% in control conditions. When Ang II was blocked with combined angiotensin receptor blockade and ACE inhibition, the decrease in SNGFR and plasma flow in response to RNS was muted. In control conditions after RNS, AR increased by approximately 50% while ER increased by about 20%, whereas after Ang II blockade, the increase in the resistances were modest (about 10% from baseline). These data suggest that the majority of renal vasoconstriction mediated by adrenergic nerve stimulation is mediated via Ang II. Interestingly, the effect of RNS on SNGFR was not altered qualitatively or quantitatively on different salt intakes, while the intrarenal Ang II content and activity should differ markedly on differing salt intakes [15].

The major adrenergic receptor subtypes to consider in the responses described above are β-, α_1- and α_2-adrenoreceptors. Certainly, adrenergic stimulation can generate Ang II by activating renin within the kidney. Our earlier studies have demonstrated that β-adrenergic receptor activation by the infusion of the β-agonist isoproterenol resulted in a reduction in LpA, with an impact on SNGFR [16, 17]. This effect on LpA was eliminated by concurrent blockade of Ang II, suggesting that β-agonist effects were mediated via activation of renin, which leads to Ang II generation in the kidney. However, nephron plasma flow was not reduced during the infusion of the β-agonist, suggesting a role for the other adrenergic receptor subtypes in the interaction with Ang II.

We have also examined the effects of α_2-agonists [18, 19]. While at lower concentrations these agents exert a capacity to suppress renin, our studies suggest that α_2-agonists act synergistically with Ang II to influence SNGFR. The α_2-agonist BHT, when infused, caused a major reduction in SNGFR via reductions in both nephron plasma flow and LpA. When Ang II receptor blockers were infused concurrently with BHT, the effects on glomerular hemodynamics and SNGFR were eliminated. A synergistic effect of Ang II infusion and BHT was also observed when the concurrent infusion of the two significantly decreased LpA (by 70%) compared to the decrease in LpA (by 50%) with Ang II alone, likely via actions on smooth muscle and glomerular cells [19]. α_1-Agonists do not appear to contribute to the interaction between Ang II and adrenergic nerve stimulation. In fact, α_1-agonists can elevate systemic blood pressure by at least 20 mm Hg without any major changes in glomerular hemodynamics [20]. In summary, renal adrenergic stimulation mediates much of the glomerular hemodynamic effects via interaction with Ang II via β-agonist stimulation of renin and cellular interactions of α_2-receptors and Ang II at the end-organ level.

The impact of multiple vasoconstrictors on glomerular hemodynamics discussed above is buffered strongly by endogenous vasodilator substances, particularly NO and prostaglandins. We have examined the effects of acute nonselective NO synthase (NOS) blockade on glomerular hemodynamics and observed that NOS blockade lowered SNGFR via reductions in nephron plasma flow and LpA, while ΔP was increased [21]. These effects were nearly identical to those of Ang II infusion [13]. As one might have predicted, concurrent application of Ang II receptor blockers eliminated the hemodynamic effects of NOS blockade. The interactions between Ang II and NO are reciprocal, wherein NO influences renin, directly and indirectly, and NO generation can be regulated by Ang II. NO also impacts glomerular hemodynamics via renal sympathetic nerves, as the hemodynamic effects of nonselective NOS blockade can be prevented by renal denervation [22]. The NO effects appear to be mediated via the α_2-adrenoreceptor, since yohimbine, an α_2-antagonist, produces the same effects as denervation. After denervation, infusion of an α_2-agonist restores the normal response to NOS blockade [22, 23]. In effect, NO acts in the kidney to buffer out the impacts of both Ang II and adrenergic stimulation.

The role of NO in glomerular ultrafiltration is also demonstrated in the specific prerenal condition associated with sepsis. Sepsis is a common predisposing clinical entity leading to prerenal AKI, often a prodrome for major tubular damage and parenchymal AKI [24–27]. Sepsis is associated with high levels of NO generated from inducible NOS (iNOS) activation. High NO levels generated by iNOS are thought to lead to autoinhibition of endothelial NOS in the kidney, thereby leaving the vasoconstrictor effects of Ang II and renal nerves relatively unopposed. Hence, in early sepsis, SNGFR decreases even before a reduction in systemic blood pressure; when we selectively inhibited iNOS, SNGFR was restored to normal values.

Prostaglandins, another class of vasodilators, also act to buffer out the hemodynamic effects of Ang II and RNS. Interactions between Ang II and prostaglandins are not simple since prostaglandins inhibit the effects of Ang II on acute administration; however, on subacute administration, certain prostaglandins act to stimulate renin and Ang II generation [28]. Inhibition of prostaglandin generation by subacute administration of cyclic oxygenase inhibitors leads to a low renin and Ang II generation and lesser vasoconstrictor influence, but acute cyclic oxygenase inhibition allows unopposed Ang II and adrenergic nerves mediated renal hemodynamic changes and reductions in GFR.

The last prerenal mechanism to consider is derived from tubular injury or reduced capacity for tubular reabsorption. A reduction in proximal tubular reabsorption increases the delivery of NaCl to the macula densa cells and activates the tubuloglomerular feedback (TGF) system, which lowers the SNGFR by afferent arteriolar vasoconstriction [29]. The role of TGF in various forms of AKI has been demonstrated before. We have shown a major role of TGF in a single-nephron model of AKI by infusing uranium, a heavy metal, into the urinary space of surface nephrons [30]. This led to a decrease in proximal tubular reabsorption and caused an immediate reduction in SNGFR. Infusion of either bumetanide or furosemide into the tubular lumina, which inhibits TGF response, prevented the reduction in SNGFR without influencing the reduction in proximal reabsorption, suggesting these effects were solely mediated via TGF.

Conclusion

Prerenal mechanisms commonly contribute to AKI in the clinical setting, particularly during major volume deficits [8–10], congestive heart failure [3], liver failure and sepsis [24, 25]. Tubular injury can also activate prerenal mechanisms via the TGF [29, 30]. The reductions in SNGFR are largely mediated by reductions in nephron plasma flow and decreases in the glomerular ultrafiltration coefficient. Adrenergic activity and Ang II are the dominant vasoconstrictors which independently and synergistically impact the determinants of glomerular filtration [10, 14, 19]. NO and prostaglandins are the major vasodilators and have complex interactions with the vasoconstrictors, but overall buffer their effects on glomerular hemodynamics [21, 22, 28]. Understanding these prerenal mechanisms which have an important role in mediating the reduction in GFR in various forms of AKI is important for the discovery and application of effective therapeutic approaches in patients with AKI.

Acknowledgements

Several of the studies described herein were supported by grants provided from the National Institutes of Health (RO-1 DK 28602 and P30-DK 079337) and funds supplied by the Department of Veterans Affairs Research Service. Dr. Prabhleen Singh was supported by a career development award from the National Institutes of Health (K08-DK-084305).

References

1 Thurau K, Boylan: Acute renal success. The unexpected logic of oliguria in acute renal failure. Am J Med 1976;61:308–15.
2 Blantz RC: Pathophysiology of pre-renal azotemia. Kidney Int 1998;53:512–523.
3 Badr KF, Ichikawa I: Prerenal failure: a deleterious shift from renal compensation to decompensation. N Engl J Med 1988;319:623–629.
4 Tucker BJ, Blantz RC: An analysis of the determinants of nephron filtration rate. Am J Physiol 1977;232:F477–F483.
5 Blantz RC: The mechanism of acute renal failure after uranyl nitrate. J Clin Invest 1975;55:621–635.
6 Blantz RC: Dynamics of glomerular ultrafiltration in the rat. Fed Proc 1977;36:2602–2608.
7 Blantz RC, Rector FC Jr, Seldin DW: The effect of hyper-oncotic albumin expansion on glomerular ultrafiltration in the rat. Kidney Int 1974;6:209–221.
8 Steiner RW, Tucker BJ, Blantz RC: Glomerular hemodynamics in rats with chronic sodium depletion: effect of saralasin. J Clin Invest 1979;64:503–512.
9 Tucker BJ, Blantz RC: Mechanism of altered glomerular hemodynamics during chronic sodium depletion. Am J Physiol 1983;244:F11–F18.
10 Tucker BJ, Mundy CA, Blantz RC: Adrenergic and angiotensin II influences on renal vascular tone in chronic sodium depletion. Am J Physiol 1987;252:F811–F817.

11 Thomson SC, Tucker BJ, Gabbai FB, Blantz RC: Functional effects on glomerular hemodynamics of short-term chronic cyclosporine in male rats. J Clin Invest 1989;83:960–969.
12 De Nicola L, Thomson SC, Wead LM, Brown MR, Gabbai FB: Arginine feeding modifies cyclosporine nephrotoxicity in rat. J Clin Invest 1993;92:1859–1865.
13 Blantz RC, Konnen KS, Tucker BJ: Angiotensin II effects upon the glomerular microcirculation and ultrafiltration coefficient of the rat. J Clin Invest 1976;57:419–434.
14 Pelayo JC, Ziegler MG, Blantz RC: Angiotensin II in adrenergic-induced alterations in glomerular hemodynamics. Am J Physiol 1984;247:F799–F807.
15 Tucker BJ, Peterson PW, Munger L, Bird E, Mitchell M, Pelayo JC, Blantz RC: Glomerular hemodynamic alterations during renal nerve stimulation in rats on high and low salt diets. Am J Physiol 1990;258:F133–F143.
16 Pelayo JC, Tucker BJ, Blantz RC: Effects of beta-adrenergic stimulation with isoproterenol upon glomerular hemodynamics. Am J Physiol 1989;257:F866–F873.
17 Thomson SC, Tucker BJ, Blantz RC: Effects of beta-adrenergic blockade on the glomerular and tubular response to acute renal denervation. J Cardiovasc Pharmacol 1991;17:397–402.
18 Thomson SC, Tucker BJ, Gabbai FB, Blantz RC: Glomerular hemodynamics and alpha-2 adrenoreceptor stimulation: the role of renal nerves. Am J Physiol 1990;258:F21–F27.
19 Thomson SC, Gabbai FB, Tucker BJ, Blantz RC: Interaction between α_2-adrenergic and angiotensin II systems in the control of glomerular hemodynamics as assessed by renal micropuncture in the rat. J Clin Invest 1992;90:604–611.
20 Pelayo JC, Tucker BJ, Blantz RC: Differential effects of methoxamine (MTX) and isoproterenol (ISO) on glomerular hemodynamics. Kidney Int 1983;23:247.
21 De Nicola, Blantz RC, Gabbai FB: Nitric oxide and angiotensin II. Glomerular and tubular interaction in the rat. J Clin Invest 1992;89:1248–1256.
22 Gabbai FB, Thomson SC, Peterson O, Wead L, Malvey K, Blantz RC: Glomerular and tubular interactions between renal adrenergic activity and nitric oxide. Am J Physiology 1995;37:F1004–F1008.
23 Vallon V, Peterson OW, Gabbai FB, Blantz RC, Thomson SC: Interactive control of renal function by alpha 2-adrenergic system and nitric oxide: role of angiotensin II. J Cardiovasc Pharmacol 1995;26:916–922.
24 Schwartz D, Mendonca M, Schwartz I, Xia Y, Satriano J, Wilson CB, Blantz RC: Inhibition of constitutive nitric oxide synthase (NOS) by nitric oxide generated by inducible NOS after lipopolysaccharide administration, provokes renal dysfunction in rats. J Clin Invest 1997;100:439–448.
25 Schwartz D, Blantz RC: Nitric oxide, sepsis, and the kidney. Semin Nephrol 1999;19:272–276.
26 Sade K, Schwartz D, Wolman Y, Schwartz I, Chernichovski T, Blum M, Brazowski E, Raz I, Blantz RC, Iaina A: Time course of lipopolysaccharide-induced nitric oxide synthase mRNA expression in rat glomeruli. J Lab Clin Med 1999;134:471–477.
27 Satriano J, Schwartz D, Ishizuka S, Lortie MJ, Thomson SC, Gabbai F, Kelly CJ, Blantz RC: Suppression of inducible nitric oxide generation by agmatine aldehyde: beneficial effects in sepsis. J Cell Physiol 2001;188:313–320.
28 Tucker BJ, Blantz RC: Acute and sub-acute prostaglandin and angiotensin inhibition on glomerular and tubular dynamics in the rat. Am J Physiol 1990;258:F1026–F1035.
29 Tucker BJ, Steiner RW, Gushwa L, Blantz RC: Studies on the tubuloglomerular feedback system in the rat. Mechanism of reduction in filtration rate with benzolamide. J Clin Invest 1978;62:993–1004.
30 Peterson OW, Gabbai FB, Myers R, Mizisin A, Blantz RC: A single nephron model of acute tubular injury: role of tubuloglomerular feedback. Kidney Int 1989;36:1037–1044.

Roland C. Blantz, MD
University of California San Diego and VASDHS
Division of Nephrology-Hypertension
3350 La Jolla Village Drive (9111-H), San Diego, CA 92161-0002 (USA)
Tel. +1 858 552 7528, E-Mail rblantz@ucsd.edu

Kellum JA, Ronco C, Vincent J-L (eds): Controversies in Acute Kidney Injury.
Contrib Nephrol. Basel, Karger, 2011, vol 174, pp 12–21

The Role of Tubuloglomerular Feedback in the Pathogenesis of Acute Kidney Injury

Prabhleen Singh[a,b] · Mark D. Okusa[c]

[a]Division of Nephrology, University of California, San Diego and VA San Diego Health System, and [b]O'Brien Center for Acute Kidney Injury Research, San Diego, Calif., [c]Division of Nephrology and Center for Immunity, Inflammation and Regenerative Medicine, University of Virginia, Charlottesville, Va., USA

Abstract

The mechanisms involved in reduction in glomerular filtration rate (GFR) in prerenal and intrarenal acute kidney injury (AKI) are not mutually exclusive and prerenal mechanisms continue to play a role in the pathogenesis of established intrarenal AKI. In nearly all forms of AKI, glomeruli are morphologically normal; therefore, the investigative efforts have focused on systemic and intrarenal mechanisms that lead to the failure of filtration at the glomerulus. There is observed and/or deductive evidence supporting the role of tubuloglomerular feedback in mediating the reduction in GFR in various forms of AKI. In prerenal AKI, the activation of various neurohormonal renal vasoconstrictors can increase the sensitivity and responsiveness of tubuloglomerular feedback. In different forms of intrarenal AKI, the varying degree of tubular injury is linked to filtration failure directly by mechanisms such as tubular obstruction or tubular backleak of solutes, or indirectly via the activation of tubuloglomerular feedback. Tubular obstruction or backleak of solutes, while readily understood, do not appear to be consistent features in experimental AKI and have a limited role in explaining the degree of impairment of GFR in human AKI. The functional connection between tubular damage and filtration failure mediated by tubuloglomerular feedback via alterations in nephron plasma flow and glomerular capillary hydrostatic pressure is more consistently observed or deduced from experimental data. It also explains the principal abnormality of increased preglomerular resistances, a pathogenic characteristic of both experimental and human AKI.

Acute kidney injury (AKI) is typically categorized as prerenal to signify hemodynamic changes in the kidney without tubular injury, or as intrarenal to refer to AKI with significant tubular injury or tubular necrosis. It is important to recognize that these categories and the processes underlying their pathogenesis are not mutually exclusive. In effect, they fall along the spectrum of progressive renal dysfunction. For instance, in the common clinical scenario of sepsis-related AKI, a persistent and progressive prerenal state develops into acute tubular necrosis. On the other hand, prerenal mechanisms continue to play a role in the decline in glomerular filtration rate (GFR) even with established tubular injury. In fact, tubular injury itself leads to the activation of prerenal mechanisms such as tubuloglomerular feedback in various forms of AKI as reviewed below.

Overview of Tubuloglomerular Feedback

In biological systems like the human body, the property that maintains the stability of the internal environment is termed homeostasis. Feedback, an elementary control system, is at the core of homeostasis, and can be defined by the change in output of a system relative to the input due to any disturbance outside the system. Feedback can be *positive* or *negative,* but in physiology, negative feedback systems are ubiquitous. The tubuloglomerular feedback system in the kidney is a prototype of the negative feedback system.

The existence of the tubuloglomerular feedback system was predicted in the 1930s by Smith [1] and Goormaghtigh [2], and was later confirmed by Thurau and Schnermann [3]. Using in vivo renal micropuncture with pharmacological manipulation in rats and later in genetically manipulated mouse models, major advances in our understanding of the tubuloglomerular feedback system have been made. Tubuloglomerular feedback operates within each nephron principally through the juxtaglomerular apparatus and renders the single-nephron GFR (SNGFR) inversely dependent on the tubular flow rate past the macula densa. Tubuloglomerular feedback acts to stabilize the nephron function and distal delivery of solutes. The structural construct of tubuloglomerular feedback is provided by the anatomical proximity of the macula densa region of the thick ascending limb of Henle's loop and the afferent arteriole of the glomerulus.

The primary signal initiating the tubuloglomerular feedback response is luminal NaCl entry via the Na,K,2Cl cotransporter into the macula densa cells. Predictably, tubuloglomerular feedback is attenuated when transport is blocked through this channel with furosemide [4] and in mice deficient in the Na,K,2Cl cotransporter [5]. While the intervening steps remain uncertain, it is generally agreed that there is a release of ATP from the macula densa cells on the basolateral side. Several lines of evidence support adenosine, generated from the released ATP by the enzyme 5′nucleotidase, as the final mediator of the vasoconstrictor response in the afferent arteriole via adenosine 1 receptors [6]. Recent advances in the underlying mechanisms of tubuloglomerular feedback have recently been reviewed in detail [7, 8] (fig. 1).

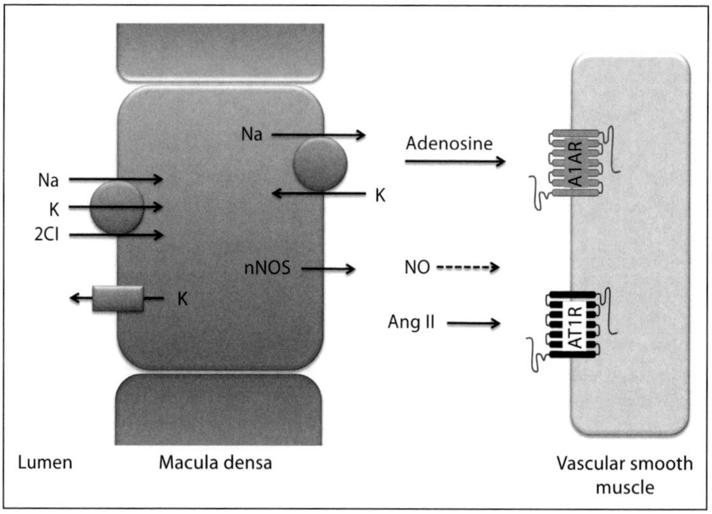

Fig. 1. Tubuloglomerular feedback, sensing and effector mechanism. NaCl is sensed by the furosemide-sensitive Na,K,2Cl cotransporter (NKCC2B and NKCC2A), and the ROMK channel mediates K efflux through the apical membrane. NaCl absorption leads to adenosine production and activation of adenosine 1A receptors (A1AR) mediating contraction of vascular smooth muscle. Solid arrows indicate positive/stimulatory relationships, and broken arrows indicate negative/inhibitory relationships. Angiotensin II (Ang II) and nitric oxide (NO) produced by neuronal NO modulates the tubuloglomerular feedback response. Adapted from Schnermann and Briggs [7].

Role of Tubuloglomerular Feedback in the Pathogenesis of Acute Kidney Injury

The connection between tubular injury with a reduced reabsorptive capacity and a consequent fall in GFR was first made by Brandt-Rehberg in 1929, but, it was Thurau and Boylan [9] who theorized that tubuloglomerular feedback dictates this relationship between reduced tubular reabsorptive capacity and reduction in GFR in AKI. In an elegant paper, they postulated that oliguria in AKI occurred as a functional adaptation by the kidney due to tubuloglomerular feedback-mediated reduction in GFR to defend against losses in extracellular volume when proximal reabsorption was compromised due to tubular injury. As more than 90% of filtered salt and water is typically reabsorbed, they hypothesized that without the presence of tubuloglomerular feedback, even a modest decrease in proximal reabsorption would lead to inordinate losses of electrolytes and water (fig. 2). They termed this immediate reduction of GFR through tubuloglomerular feedback to minimize the loss of extracellular volume until tubule function is reestablished 'acute renal success' instead of acute renal failure.

In various forms of AKI, either a decrease in proximal reabsorption (due to lethal and sublethal tubular injury in intrarenal AKI) leading to

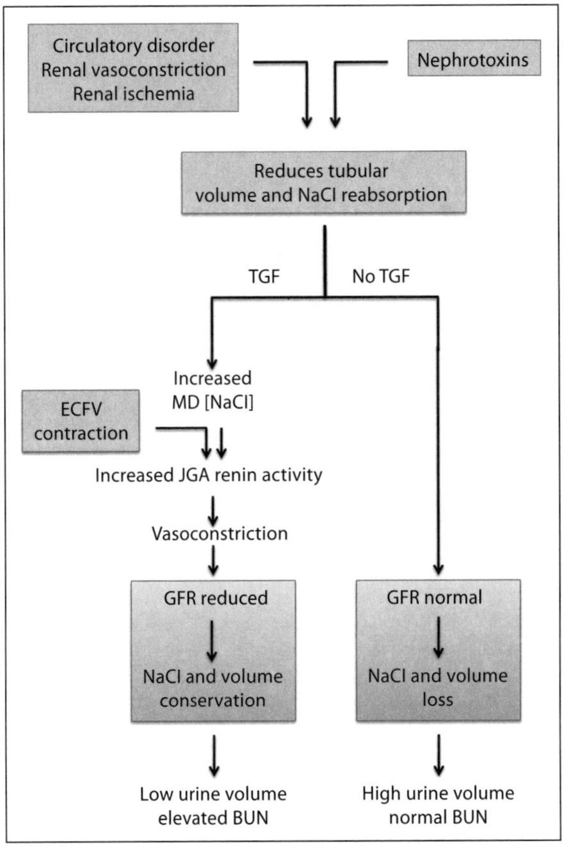

Fig. 2. Postulated effect of tubuloglomerular feedback (TGF) activation on urine volume and renal function in AKI. In ischemic or nephrotoxic tubular injury, reduced tubular reabsorption of NaCl in proximal tubule segments leads to an increase in macula densa (MD) NaCl. The increase in signal at the macula densa then mediates vasoconstriction of the afferent arteriole and reduction of GFR via tubuloglomerular feedback. The tubuloglomerular feedback-mediated reduction in GFR prevents the loss of extracellular volume and sodium, but at the expense of retention of nitrogenous wastes. ECFV = Extracellular fluid volume; JGA = juxtaglomerular apparatus; BUN = blood urea nitrogen. Adapted from Thurau and Boylan [9].

tubuloglomerular feedback activation or an increased responsiveness of the tubuloglomerular feedback system itself (in prerenal AKI) can mediate the decline in SNGFR. In prerenal AKI, the activation of various neurohormones in response to 'effective' circulating volume influences the physical determinants of filtration. In the early stages of prerenal azotemia, the actions of neurohormones such as angiotensin II (Ang II) and catecholamines lead to an increase in proximal tubule sodium reabsorption. This should decrease the signal for tubuloglomerular feedback activation and cause a vasodilatory response and increase in SNGFR. However, the same neurohormones, specifically Ang II, concomitantly increase tubuloglomerular feedback responsiveness, such that the net influence of tubuloglomerular feedback on SNGFR is maintained [10]. The increased responsiveness of the tubuloglomerular feedback system implies that the system is engaged at a lower rate of tubular flow past the macula densa. There is evidence to support that the enhanced tubuloglomerular feedback responsiveness due to Ang II occurs independently of the direct vasoconstrictor effects of Ang II [10].

It is pertinent to note that in most forms of intrarenal AKI, the glomeruli are morphologically normal, but tubular injury is consistent, although variable in degree. Hence, the investigative focus has centered on tubular injury, but it is equally important to focus on the mechanisms that link tubular injury to the failure of filtration at the glomerulus. Renal micropuncture studies have been elemental in determining the various pathogenic factors that connect tubular factors to alterations in glomerular filtration. The physical factors that could mediate a primary failure of filtration include nephron plasma flow, hydrostatic pressure gradient (ΔP) between glomerular capillary and Bowman's space, and glomerular ultrafiltration coefficient (LpA). The contributions of these factors vary in different forms or models of AKI and there has been much debate on their relative significance in the pathogenesis of AKI [11].

Proximal tubular injury with impaired reabsorptive capacity is a uniform feature in various forms of AKI. Tubular injury can lead to a decrease in GFR directly by mechanical factors such as tubular obstruction with epithelial debris leading to increased tubular pressures which lower the net driving force for filtration (ΔP). However, tubular obstruction does not appear to be as extensive in human studies compared to rodent models, and it does not appear to be a significant pathogenic mechanism in AKI in humans [10, 12]. Decreased clearance of solutes can also occur without a primary failure in filtration with transepithelial backleak of solutes across damaged tubular epithelium due to cytoskeletal breakdown. The degree of extensive tubular damage observed in animal models that demonstrate backleak is rarely observed in AKI in humans [13]. Human studies have demonstrated that tubular backleak cannot singularly account for the overall decrease in clearance as only a 10% decrease in GFR could be explained by tubular backleak [14, 15].

Based on the available data, the major prerenal mechanism linking proximal tubular injury to filtration failure at the glomerulus appears to be tubuloglomerular feedback via its control on the preglomerular resistance. Applying network modeling to glomerular hemodynamics, Oken [16] demonstrated that a predominant increase in preglomerular resistances, with or without a change in postglomerular resistances, can account for both the cortical ischemia and filtration failure typically observed in human AKI. When tubuloglomerular feedback is examined at a single-nephron level by perturbations in the late proximal flow, a 30–60% decline in SNGFR has been observed [17, 18]. This suggests that tubuloglomerular feedback alone could not explain the fall in GFR to less than

10% of normal frequently observed in AKI. However, the aggregate response to a simultaneous activation of tubuloglomerular feedback in all nephrons within a kidney appears to be greater than the sum of the responses if each nephron were activated separately [17]. In addition, persistent preglomerular vasoconstriction and increased sensitivity to vasoconstrictors observed in ischemic kidneys can increase the response of the tubuloglomerular feedback system in AKI [12, 19, 20]. Finally, an abnormal renal autoregulatory response has been observed in ischemic kidneys where a decrease in renal pressure leads to renal vasoconstriction instead of vasodilation [21]. The direct and indirect evidence supporting tubuloglomerular feedback-mediated vasoconstriction leading to a reduction in GFR in different etiologies of AKI is reviewed below.

Ischemic AKI

In ischemic AKI, the duration of ischemia appears to influence the factors that mediate the decrease in GFR. In severe ischemic injury, such as after an hour of cessation of renal blood supply, tubular obstruction assumes a more important role as a cause of filtration failure. However, significant heterogeneity in tubular pressures has been observed with normal pressures in several tubules, making it unlikely for tubular obstruction to be the singularly important mediator of reduction in GFR after ischemic injury [22]. Arendshorst et al. [23] in 1975 studied severe AKI after 1 h of unilateral renal ischemia. They found that in the early stages (1–3 h) after ischemia, the pressures in the proximal and distal tubules were heterogeneous but elevated, and there was histologic evidence of tubular obstruction. Although renal blood flow was reduced and renal vascular resistances were increased, glomerular capillary pressure was normal, suggesting parallel increases in both afferent and efferent arteriolar resistances. Tubular obstruction appeared to be the predominant factor in the decline in filtration rate at this stage. However, at a later time-point (22–26 h), tubular pressures were normal or subnormal, but preglomerular vasoconstriction was marked, likely due to activation of tubuloglomerular feedback, and this was thought to contribute to the persistent decline in filtration rate.

Mason et al. [24] in 1977 similarly found increased tubular pressures in the early stages at 1–4 h after either 60 or 75 min of unilateral ischemia. However, with milder ischemia of 45-min duration, the proximal tubules were not dilated with more than 75% of the tubular pressures in the normal range, suggesting that other mechanisms mediated the decline in GFR. Next, the role of tubuloglomerular feedback in the initial phase (1–4 h) after 45- or 75-min ischemia was examined by them [25]. Early proximal flow rates were measured during different flow conditions in the loop of Henle. These conditions included no flow, orthograde perfusion with Ringer's solution (tubuloglomerular feedback activation), orthograde perfusion with Ringer's solution and furosemide (tubuloglomerular feedback inhibition), and retrograde perfusion with mannitol (to dilute the NaCl concentration, inhibit tubuloglomerular feedback) and Ringer's solution to activate tubuloglomerular feedback. Normal tubuloglomerular feedback responses were observed in both durations of ischemia experiments, demonstrating a viable tubuloglomerular feedback system in acutely injured kidneys. Importantly, they assessed whether the lower limit of tubuloglomerular feedback-mediated depression in filtration rate was sufficient to account for the reduction in whole kidney GFR in AKI, and found that the whole kidney GFR was lowered to a degree compatible with the operating range of feedback response at a single-nephron level during both periods of ischemia.

Nephrotoxic Acute Kidney Injury

Mason et al. [25] found a viable tubuloglomerular feedback system after other forms of renal injury as well, such as nephrotoxic and pigment nephropathy, with varying degrees of contribution to the decline in GFR. Blantz et al. [26] also examined and found a pathogenic role of tubuloglomerular feedback in nephrotoxic models of AKI. In a systemic uranyl nitrate model, a decline in ultrafiltration coefficient (LpA) was observed along with renal tubular damage leading to tubular obstruction, but this was not prominent [26]. Importantly, the reduction in LpA appeared to be functional in nature, as with plasma volume expansion or captopril administration, and the effects on LpA and SNGFR were no longer observed. This suggested that despite the structural damage, functional reductions in GFR could be independently altered by maneuvers that diminish the tubuloglomerular feedback system (such as volume expansion and angiotensin inhibition). More importantly, a direct and major role for tubuloglomerular feedback has been demonstrated by studies in a single-nephron model of AKI [27]. Within minutes of intratubular administration of uranyl nitrate, a 16–30% reduction in SNGFR was observed. As there was no systemic exposure to the toxin, only a local mechanism such as tubuloglomerular feedback could be invoked to explain the immediate decline in SNGFR. Moreover, when tubuloglomerular feedback was inhibited with high doses of intratubular furosemide and bumetanide microperfusion, the reduction in SNGFR was prevented. Finally, micropuncture experiments in the uranyl nitrate model of AKI by another group also found altered renal hemodynamics due to tubuloglomerular feedback activation to be the primary mediator of GFR decline in the early and late stages after exposure to the toxin [28]. In other nephrotoxic models, such as those due to mercuric chloride or aminoglycoside administration, although proximal tubular injury and necrosis is quite evident, a failure of filtration was thought to be the primary abnormality, with secondary development of tubular obstruction as a result of reduced filtration rate [11, 15, 29]. Similar to uranyl nitrate models, measures such as salt loading or angiotensin blockade that alter renal hemodynamics and suppress the tubuloglomerular feedback system are protective [11, 15].

In AKI associated with increased filtered load of endogenous substances such as hyperuricemia or rhabdomyolysis, clinically or experimentally with glycerol administration, tubular obstruction has been demonstrated by many to be the primary mechanism leading to renal failure. Others have argued a primary filtration failure as the cause of tubular collapse and obstruction rather than the other way around [11]. The finding of low intratubular pressures in obstructed tubules without any evidence of major tubular backleak of solutes explaining the dissipation of high intratubular pressures appear to support this hypothesis. The idea is further supported by the maintenance of filtration with maneuvers that may influence preglomerular resistances. While a direct role of tubuloglomerular feedback has not been examined in these conditions, a significant decrease in renal plasma flow also appeared to be important in the pathogenesis [15]. Specifically, in experiments utilizing methemoglobin to induce AKI, in addition to widespread tubular collapse with low intratubular pressure, decreased efferent arteriolar pressures were found, and an increase in preglomerular resistance was invoked to explain the pathogenesis of AKI [30].

Conclusion

In summary, the significance of certain factors such as tubular obstruction, reduction in LpA and transepithelial tubular backleak of filtrate in the pathogenesis of AKI appears to be dependent on the model of AKI utilized in a study. More often than not, their significance in the pathogenesis of AKI appears to be alterable by measures that primarily influence renal hemodynamics. Moreover, with respect to AKI observed clinically, the contribution of tubular obstruction or backleak of filtrate appears to be minimal in milder cases of AKI [16]; however, in more severe forms of AKI and later in the time course of AKI, tubular obstruction and backleak of filtrate may become increasingly important [23]. An increase in preglomerular resistance appears to be consistent in nearly all forms of AKI, hence the popularity of the term 'vasomotor nephropathy' to describe AKI. A direct or indirect role of tubuloglomerular feedback in mediating this cardinal renal hemodynamic alteration has been observed in nearly all forms of AKI.

Future Direction

The recent focus on prerenal azotemia in conditions such as cardiorenal syndrome (CRS) has raised additional questions about the role of tubuloglomerular feedback in the pathogenesis of AKI. Five forms of CRS have been defined recently [31]. Of these, type 1 CRS, in which acute cardiac events such as acute diastolic heart failure lead to AKI, and type 2 CRS, in which chronic heart disease such as chronic heart failure is associated with both an increased susceptibility to AKI as well as progressive CKD, appear particularly relevant. CRS types 1 and 2 are thought to mediate AKI via sustained actions of the renin-angiotensin-aldosterone system and sympathetic nervous system, systemic oxidative stress, volume overload, hypertension, and systemic inflammation. Of these, factors such as the renin-angiotensin system, adrenergic nerves and adenosine potentiate tubuloglomerular feedback. In addition, in CRS there is impairment in the production of local vasodilators such as nitric oxide (endothelial dysfunction) and prostaglandins, which can further activate tubuloglomerular feedback. This suggests that tubuloglomerular feedback activity in CRS is likely to be intensified. But, given the complex systemic and intrarenal hemodynamic alterations in CRS, the specific role of tubuloglomerular feedback has not been determined. Furthermore, how an increase in central venous pressure from congestive heart failure affects tubuloglomerular feedback remains to be elucidated. These questions will need to be addressed in future studies in order to fully understand the role of tubuloglomerular feedback in mechanisms of cardiorenal syndrome.

Acknowledgements

Dr. Prabhleen Singh is supported by a career development award from the National Institutes of Health (K08-DK-084305). Dr. Mark Okusa is supported from the National Institutes of Health (R01 DK062324, R01 DK0834061, and R01 DK085259).

References

1 Smith HW: Kidney. Annu Rev Physiol 1939;1:503–528.
2 Goormaghtigh N, L'appareil neuro-myo-arte´riel juxtaglomerulaire du rein; ses reactions en pathologie et ses rapports avec le tube urinifere. C R Seances Soc Biol Fil 1937;124:293–296.
3 Thurau K, Schnermann J: Die Natriumkonzentration an den Macula Densa-Zellen als regulierender Faktor für das Glomerulumfiltrat (Mikropunktionsversuche). Klin Wochenschr 1965;43:410–413.
4 Wright FS, Schnermann J: Interference with feedback control of glomerular filtration rate by furosemide, triflocin, and cyanide. J Clin Invest 1974;53:1695–1708.
5 Oppermann M, et al: Renal function in mice with targeted disruption of the A isoform of the Na-K-2Cl co-transporter. J Am Soc Nephrol 2007;18:440–448.
6 Vallon V, Muhlbauer B, Osswald H: Adenosine and kidney function. Physiol Rev 2006;86:901–940.
7 Schnermann J, Briggs JP: Tubuloglomerular feedback: mechanistic insights from gene-manipulated mice. Kidney Int 2008;74:418–426.
8 Singh P, Thomson SC: Renal homeostasis and tubuloglomerular feedback. Curr Opin Nephrol Hypertens 2010;19:59–64.
9 Thurau K, Boylan JW: Acute renal success. The unexpected logic of oliguria in acute renal failure. Am J Med 1976;61:308–315.
10 Braam B, et al: Relevance of the tubuloglomerular feedback mechanism in pathophysiology. J Am Soc Nephrol 1993;4:1257–1274.
11 Oken DE: Acute renal failure (vasomotor nephropathy): micropuncture studies of the pathogenetic mechanisms. Annu Rev Med 1975;26:307–319.
12 Schrier RW, et al: Acute renal failure: definitions, diagnosis, pathogenesis, and therapy. J Clin Invest 2004;114:5–14.
13 Edelstein C, Schrier RW: Pathophysiology of ischemic acute renal failure; in Schrier RW (ed): Diseases of the Kidney and Urinary Tract. Philadelphia, Lippincot Williams & Wilkins, 2001, pp 1041–1070.

14 Myers BD, et al: Transtubular leakage of glomerular filtrate in human acute renal failure. Am J Physiol 1979;237:F319–F325.
15 Bird JE, Blantz RC: Acute renal failure: the glomerular and tubular connection. Pediatr Nephrol 1987;1:348–358.
16 Oken DE: Hemodynamic basis for human acute renal failure (vasomotor nephropathy). Am J Med 1984;76:702–710.
17 Blantz RC, et al: Physiologic adaptations of the tubuloglomerular feedback system. Kidney Int 1990;38:577–583.
18 Schnermann J, et al: Regulation of superficial nephron filtration rate by tubulo-glomerular feedback. Pflugers Arch 1970;318:147–175.
19 Bonventre JV, Weinberg JM: Recent advances in the pathophysiology of ischemic acute renal failure. J Am Soc Nephrol 2003;14:2199–2210.
20 Conger JD, Weil JV: Abnormal vascular function following ischemia-reperfusion injury. J Investig Med 1995;43:431–442.
21 Conger JD, Robinette JB, Schrier RW: Smooth muscle calcium and endothelium-derived relaxing factor in the abnormal vascular responses of acute renal failure. J Clin Invest 1988;82:532–537.
22 Conger JD, Robinette JB, Kelleher SP: Nephron heterogeneity in ischemic acute renal failure. Kidney Int 1984;26:422–429.
23 Arendshorst WJ, Finn WF, Gottschalk CW: Pathogenesis of acute renal failure following temporary renal ischemia in the rat. Circ Res 1975;37:558–568.
24 Mason J, et al: The early phase of experimental acute renal failure. I. Intratubular pressure and obstruction. Pflugers Arch 1977;370:155–163.
25 Mason J, et al: The early phase of experimental acute renal failure. III. Tubologlomerular feedback. Pflugers Arch 1978;373:69–76.
26 Blantz RC, et al: The mechanism of acute renal failure after uranyl nitrate. J Clin Invest 1975;55:621–635.
27 Peterson OW, et al: A single nephron model of acute tubular injury: role of tubuloglomerular feedback. Kidney Int 1989;36:1037–1044.
28 Flamenbaum W, et al: Uranyl nitrate-induced acute renal failure in the rat: micropuncture and renal hemodynamic studies. Kidney Int 1974;6:408–418.
29 Flanigan WJ, Oken DE: Renal micropuncture study of the development of anuria in the rat with mercury-induced acute renal failure. J Clin Invest 1965;44:449–457.
30 Ruiz-Guinazu A, Coelho JB, Paz RA: Methemoglobin-induced acute renal failure in the rat. In vivo observation, histology and micropuncture measurements of intratubular and postglomerular vascular pressures. Nephron 1967;4:257–275.
31 Ronco C, et al: Cardiorenal syndromes: an executive summary from the consensus conference of the Acute Dialysis Quality Initiative (ADQI); in Ronco C, Bellomo R, McCullough PA (eds): Cardiorenal Syndromes in Critical Care. Contrib Nephrol. Basel, Karger, vol 165, pp 54–67.

Prabhleen Singh, MD
Assistant Professor of Medicine, Division of Nephrology
University of California, San Diego & VASDHS, Division of Nephrology-Hypertension
3350 La Jolla Village Drive (9111-H), San Diego, CA 92161–0002 (USA)
Tel. +1 858 552 8585, ext 2609, E-Mail p1singh@ucsd.edu

Kellum JA, Ronco C, Vincent J-L (eds): Controversies in Acute Kidney Injury.
Contrib Nephrol. Basel, Karger, 2011, vol 174, pp 22–32

Can We Identify Prerenal Physiology and Does It Matter?

Didier Payen · Matthieu Legrand

Department of Anesthesiology, Critical Care and SAMU, Hôpital Lariboisière, Assistance Publique-Hôpitaux de Paris, Laboratoire d'Anesthésiologie, Université Paris, Paris, France

Abstract

The classic concept of 'prerenal azotemia' is an entity not well diagnosed in the clinic as it is mainly the time evolution that confirms the absence of renal injury. However, this entity seems to be associated with a worse prognosis compared to patients with normal kidney function. In intensive care unit (ICU) patients, this entity results from the interaction with outside factors, renal hypoxia and cellular infiltration by immune cells. The mechanism of such an entity may result from perfusion abnormalities, but more importantly from an oxygenation deficit. This syndrome can be seen as the first step in a continuum from adapted renal function to the occurrence of renal injury. Early renal hemodynamics might be important for predicting the occurrence of AKI. However, the paradigm of renal ischemia as a major mechanism is associated with microcirculation alteration and immune cell infiltration in the generation of AKI. Because creatinine elevation is delayed from renal injury, early detection might help in deciding on the therapeutic strategy. To achieve such a goal, the development of biomarkers for renal injury might be helpful. Some molecules such as NGAL, kidney injury molecule-1, interleukin-18 and cystatin have been proposed and validated as predictors of renal injury in clinical contexts. Although controversial results have been published, most of the results demonstrate a normal level in the presence of prerenal azotemia compared to AKI, limiting their interest for prediction. Importantly, clusters of proteome in urine might improve the sensitivity and specificity to predict AKI in presence of prerenal azotemia.

Kidney functions interact with many systems and control the whole body fluid balance, electrolytes, acid-base status and size of fluid compartments as extracellular, intracellular and vascular volumes. In addition to its intrinsic specificities, the kidney is a major part for many regulatory systems ('outside' renal factors) that may change kidney condition and function (fig. 1). Among the

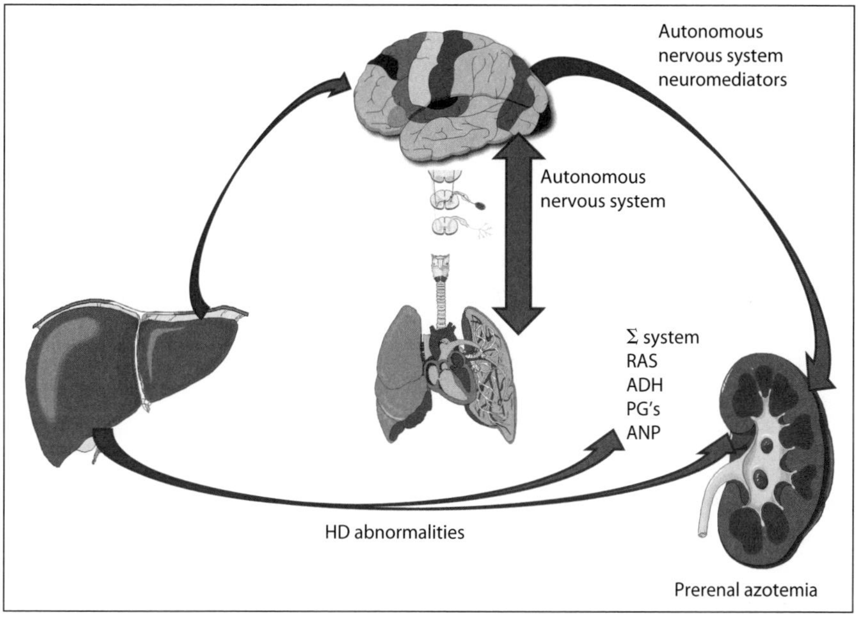

Fig. 1. Interaction between 'outside factors' and kidney function.

conditions which have renal function consequences, renal circulation, oxygenation and inflammation are the most frequent in intensive care unit (ICU) patients. Conversely, parenchymal kidney disease as a comorbidity may limit the homeostatic renal functions, altering the nonlimited kidney capabilities to correct an acute change in pH, sodium balance, blood pressure, etc. An acute situation in critically ill patients frequently associates perfusion abnormalities (hypotension, hypovolemia, reduced renal blood flow), endothelial dysfunction (abnormal permeability, modified function for endothelial receptors, procoagulation status), stimulation of regulatory systems (autonomic nervous system, renin-angiotensin system, aldosterone, ADH). Intrarenal regulation might also interfere with the above-mentioned systems, changing vascular resistance in pre- and postglomerular vessels, tubuloglomerular feedback, and kidney cell energy. The concept of 'prerenal azotemia' in this context has to be clarified and remains a difficult entity to diagnose.

The Classic Paradigm of Prerenal Azotemia: Definitions and Concepts

Because acute renal failure (ARF) in ICU patients is associated with higher mortality [1, 2] and the increase in classic markers of renal dysfunction such as creatinine and urea are delayed from the onset of renal tissue injury, the new scoring system quoting the risk of injury in addition to injury and failure (RIFLE or AKIN) was developed [3]. It was even argued that patients die *with* acute kidney injury (AKI) rather than *of* AKI; the statistical association between AKI and high mortality does not imply a mechanistic relation. Some questions still remain unresolved, especially for the so-called 'prerenal azotemia', such as its impact on ICU patient mortality, the presence or absence of renal tissue injury, the incidence in ICU, etc.

The spectrum of AKI contains all the previous classifications of ARF, especially the one called 'prerenal failure', 'prerenal azotemia' or 'functional renal failure'. The 'prerenal' entity suggests that the kidney function is altered in relation with events 'outside' the kidney, in opposition with renal parenchyma alterations ('renal') or urinary excretory obstruction ('postrenal'). The 'prerenal' term corresponds to various intricate mechanisms similar to those observed in hepatorenal syndrome [4] or severe cardiac failure [5], modifying renal function but without primary parenchymal disease. It can be seen as an adapted renal response to these outside factors. Despite large interindividual variations, AKI can be seen as a continuum between prerenal azotemia and severe AKI, with renal 'structural' injury such as acute tubular necrosis (ATN). If this view is correct, it implies that ATN cannot arise de novo, but must be secondary to sustained or not sufficiently corrected functional injuries, progressing even to the point of cell necrosis.

The Flaws of This Paradigm

Measurements of elevated blood urea nitrogen and serum creatinine, known as azotemia, are routinely done in the ICU, and together with oliguria or anuria characterize kidney failure [6]. The presence or absence of renal parenchyma abnormalities is difficult to diagnose in clinical conditions despite the proposed blood or urinary tests, such as detection of cellular debris in urine and determination of fractional excretion of sodium or urea [7]. None of these tests have been demonstrated accurate enough to assess the presence or absence of parenchyma injury, which can only be shown on renal histology, which is usually not performed in prerenal azotemia. Finally, it is the evolution of initial AKI that allows the classification in prerenal versus renal azotemia.

Moreover, azotemia and oliguria can also be a normal response of the kidney to 'outside' renal events, such as extracellular volume depletion, or decreased renal blood flow, severe hypotension, etc. As a consequence, changes in urine output and glomerular filtration rate (GFR) are neither necessary nor sufficient for the diagnosis of renal pathology. The 'outside' kidney reason for reduced GFR implies that a greater fraction of salt and water can be absorbed, with less fluid entering the tubules creating a subsequent oliguria. For this, prerenal azotemia is a perfectly normal response. The diagnosis of a 'normal' versus 'abnormal' renal response is then difficult in the clinical context.

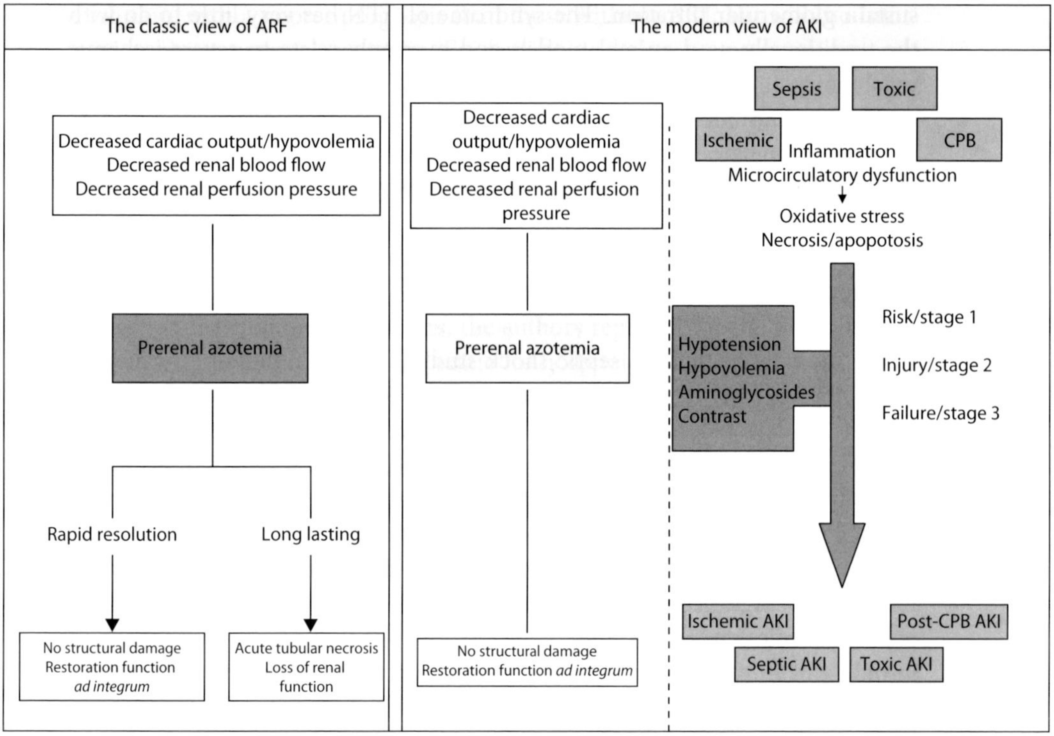

Fig. 2. Schematization of 'prerenal azotemia' positioning in the classical view of ARF and in the modern view of AKI.

The 'normal' renal response itself might put the parenchyma at risk of injury since the relationship between urine output and renal function/injury is complex. If the renal tubules work maximally to reabsorb solutes and water as a normal response to outside factors, the energetic cost of such response requires sufficient oxygen delivery. The persistence of a reduced energetic supply may then induce a part of parenchyma ischemia (fig. 2). Breaking this vicious circle requires aggressive treatment of delivery failure. As an example, oliguria may be more profound when tubular function is intact. Conversely, when tubules are injured, maximal concentrating ability is impaired and urine volume may even be normal, which calls into question the interest of azotemia plus oliguria as a hallmark of renal injury.

Acute Tubular Necrosis

ATN almost came from experimental data from a renal ischemic model showing necrotic lesions, especially on the proximal tubule [8], extended by analogy to clinical situations. In absence of histological proof, ATN is declared when there is adequate renal perfusion to largely maintain tubular integrity but not

15 Manoharan G, Pijls NH, Lameire N, Verhamme K, Heyndrickx GR, Barbato E, Wijns W, Madaric J, Tielbeele X, Bartunek J, De Bruyne B: Assessment of renal flow and flow reserve in humans. J Am Coll Cardiol 2006;47:620–625.
16 Langenberg C, Bellomo R, May C, Wan L, Egi M, Morgera S: Renal blood flow in sepsis. Crit Care 2005;9:R363–R374.
17 Lerolle N, Guerot E, Faisy C, Bornstain C, Diehl JL, Fagon JY: Renal failure in septic shock: predictive value of Doppler-based renal arterial resistive index. Intensive Care Med 2006;32:1553–1559.
18 Deruddre S, Cheisson G, Mazoit JX, Vicaut E, Benhamou D, Duranteau J: Renal arterial resistance in septic shock: effects of increasing mean arterial pressure with norepinephrine on the renal resistive index assessed with Doppler ultrasonography. Intensive Care Med 2007;33:1557–1562.
19 Darmon M, Schortgen F, Vargas F, Liazydi A, Schlemmer B, Brun-Buisson C, Brochard L: Diagnostic accuracy of Doppler renal resistive index for reversibility of acute kidney injury in critically ill patients. Intensive Care Med 2011;37:68–76.
20 Legrand M, Mik EG, Balestra GM, Lutter R, Pirracchio R, Payen D, Ince C: Fluid resuscitation does not improve renal oxygenation during hemorrhagic shock in rats. Anesthesiology 2010;112:119–127.
21 Legrand M, Mik EG, Johannes T, Payen D, Ince C: Renal hypoxia and dysoxia after reperfusion of the ischemic kidney. Mol Med 2008;14:502–516.
22 Grigoryev DN, Liu M, Hassoun HT, Cheadle C, Barnes KC, Rabb H: The local and systemic inflammatory transcriptome after acute kidney injury. J Am Soc Nephrol 2008;19:547–558.
23 Moore E, Bellomo R: Novel biomarkers of acute kidney injury: ready for clinical application? Curr Opin Crit Care 2010, E-pub ahead of print.
24 Nickolas TL, O'Rourke MJ, Yang J, Sise ME, Canetta PA, Barasch N, Buchen C, Khan F, Mori K, Giglio J, Devarajan P, Barasch J: Sensitivity and specificity of a single emergency department measurement of urinary neutrophil gelatinase-associated lipocalin for diagnosing acute kidney injury. Ann Intern Med 2008;148:810–819.
25 Parikh CR, Mishra J, Thiessen-Philbrook H, Dursun B, Ma Q, Kelly C, Dent C, Devarajan P, Edelstein CL: Urinary IL-18 is an early predictive biomarker of acute kidney injury after cardiac surgery. Kidney Int 2006;70:199–203.
26 Han WK, Bailly V, Abichandani R, Thadhani R, Bonventre JV:. Kidney Injury Molecule-1 (KIM-1): a novel biomarker for human renal proximal tubule injury. Kidney Int 2002;62:237–244.
27 Bagshaw SM, Bellomo R: Cystatin C in acute kidney injury. Curr Opin Crit Care 2010.
28 Mishra J, Dent C, Tarabishi R, Mitsnefes MM, Ma Q, Kelly C, Ruff SM, Zahedi K, Shao M, Bean J, Mori K, Barasch J, Devarajan P: Neutrophil gelatinase-associated lipocalin (NGAL) as a biomarker for acute renal injury after cardiac surgery. Lancet 2005;365:1231–1238.
29 Wagener G, Jan M, Kim M, Mori K, Barasch JM, Sladen RN, Lee HT: Association between increases in urinary neutrophil gelatinase-associated lipocalin and acute renal dysfunction after adult cardiac surgery. Anesthesiology 2006;105:485–491.
30 Herget-Rosenthal S, Marggraf G, Husing J, Goring F, Pietruck F, Janssen O, Philipp T, Kribben A: Early detection of acute renal failure by serum cystatin C. Kidney Int 2004;66:1115–1122.
31 Ahlstrom A, Tallgren M, Peltonen S, Pettila V: Evolution and predictive power of serum cystatin C in acute renal failure. Clin Nephrol 2004;62:344–350.
32 Metzger J, Kirsch T, Schiffer E, Ulger P, Mentes E, Brand K, Weissinger EM, Haubitz M, Mischak H, Herget-Rosenthal S: Urinary excretion of twenty peptides forms an early and accurate diagnostic pattern of acute kidney injury. Kidney Int 2010;78:1252–1262.

Didier M. Payen, MD, PhD
Department of Anesthesiology, Critical Care and SMUR
Hôpital Lariboisière, Assistance Publique-Hôpitaux de Paris
2 rue Ambroise-Paré, FR–75475 Paris Cedex 10 (France)
Tel. +33 1 4995 8085, E-Mail dpayen1234@orange.fr

Kellum JA, Ronco C, Vincent J-L (eds): Controversies in Acute Kidney Injury.
Contrib Nephrol. Basel, Karger, 2011, vol 174, pp 33–45

Cardiorenal Syndrome: A Complex Series of Combined Heart/Kidney Disorders

Ching Yan Goh[a,c] · Grazia Vizzi[b,c] · Massimo De Cal[b,c] · Claudio Ronco[b,c]

[a]Department of Nephrology, Selayang Hospital, Selangor, Malaysia; [b]Department of Nephrology, Dialysis & Transplantation, Ospedale San Bortolo, San Bortolo, and [c]International Renal Research Institute Vicenza (IRRIV), Vicenza, Italy

Abstract

Over the last decade, it has become increasingly clear that the cardiovascular and renal systems are interdependent. Primary disorders of either system have been shown to disturb the other system. As a result, a class of cardiorenal syndrome (CRS) has been identified wherein a vicious cycle is established in which acute/chronic dysfunction of either the kidney or the heart exacerbates the loss of function in the other organ. The ADQI organization has proposed a classification derived from a consensus conference held in 2008. CRS is classified as a disorder of the heart and kidneys whereby acute or chronic dysfunction in one organ may induce acute or chronic dysfunction in the other. The general definition has been expanded into five subtypes: CRS type 1 = acute worsening of heart function (acute heart failure-acute coronary syndrome) leading to kidney injury and/or dysfunction; CRS type 2 = chronic abnormalities in heart function (chronic heart failure-chronic heart disease) leading to kidney injury or dysfunction; CRS type 3 = acute worsening of kidney function (acute kidney injury) leading to heart injury and/or dysfunction; CRS type 4 = chronic kidney disease (chronic kidney disease) leading to heart injury, disease and/or dysfunction; and CRS type 5 = systemic conditions leading to simultaneous injury and/or dysfunction of heart and kidney. A major problem with previous terminology was that it did not allow for identification of pathophysiological interactions occurring in the different types of combined heart/kidney disorders. The subdivision into different subtypes seems to provide a better approach to this syndrome.

Cardiorenal syndromes (CRS) are broadly defined as conditions characterized by the initiation and/or progression of renal insufficiency secondary to heart failure [1], and it has also been used to describe the negative cardiac effects of declining renal function (renocardiac syndrome) [2]. A recent definition and classification system has been generated to include a vast array of acute or chronic conditions in these two important organs, where the primary failing organ can be either the heart or the kidneys [3]. Such classification has been endorsed and further developed by the ADQI (Acute Dialysis Quality Initiative) consensus group leading to a comprehensive characterization of the complex heart-kidney organ crosstalk [4]. This new classification represents a step forward to a better understanding of the pathophysiology and management strategies of the bidirectional heart-kidney interactions.

The Definition or Classification of Cardiorenal Syndromes

Any direct and indirect insult to either heart or kidneys can initiate and perpetuate the combined disorder of these two organs through several complex mechanisms. Therefore, 'cardiorenal syndromes', a large umbrella term, should be used to describe a series of heart and kidney disorders whereby acute or chronic dysfunction in one organ may induce acute or chronic dysfunction (structural or functional abnormalities) of the other. A more descriptive classification into five different subtypes would permit recognition of the primary organ dysfunction (cardiac vs. renal) as well as the acute versus chronic nature of the condition [4]. The five subtypes of the CRS can be identified (table 1) and defined as follows:

- Acute cardiorenal syndrome (CRS type 1): an acute worsening of heart function (acute heart failure-acute coronary syndrome) leading to kidney injury and/or dysfunction.
- Chronic cardiorenal syndrome (CRS type 2): chronic abnormalities in heart function (chronic heart failure-coronary heart disease) leading to kidney injury and/or dysfunction.
- Acute renocardiac syndrome (CRS type 3): acute worsening of kidney function (acute kidney injury) leading to heart injury and/or dysfunction.
- Chronic renocardiac syndrome (CRS type 4): chronic kidney disease leading to heart injury, disease and/or dysfunction.
- Secondary cardiorenal syndromes (CRS type 5): systemic conditions causing simultaneous injury and/or dysfunction of both the heart and kidneys.

The relevant mechanisms are summarized in figure 1.

Pathophysiology of Cardiorenal Syndromes

Type 1 CRS – Acute Cardiorenal Syndrome

Premorbid renal dysfunction is common in heart failure patients [de novo acute heart failure (AHF) or acute decompensated heart failure (ADHF)] and

Table 1. Modern definition of CRS

	Type 1	Type 2	Type 3	Type 4	Type 5
Organ failure sequence	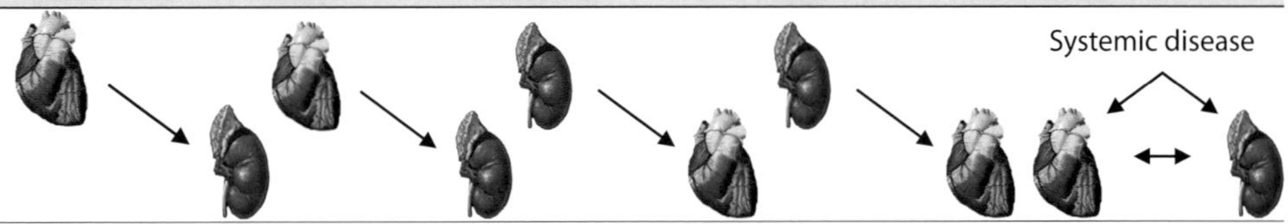				
Definition	Abrupt worsening of cardiac function leading to acute kidney injury	Chronic abnormalities in cardiac function causing progressive and permanent chronic kidney disease	Abrupt worsening of renal function causing acute cardiac disorders	CKD contributing to decreased cardiac function, cardiac hypertrophy and/ or increased risk of cardiovascular events	Systemic condition (e.g. diabetes mellitus, sepsis) causing both cardiac and renal dysfunction
Primary events	Acute decompensated heart failure, ischemic insult, coronary angio, cardiac surgery	LV remodeling and dysfunction Diastolic dysfunction Chronic abnormalities in cardiac function	AKI (e.g. acute kidney ischaemia or glomerulo-nephritis)	CKD (e.g. chronic glomerular disease)	Sepsis
Criteria for primary events	Severe arrythmia Troponin/ST elevation ECHO: dilatation	Decreased EF Increased CVP NYHF class 3–4	RIFLE R.I.F. or AKIN stage 1-2-3 or AKI requiring RRT	CKD stage 1–5 or CKD requiring RRT	APACHE 2 > SOFA SCORE > Number of failing organs
Sequelae	Inadequate renal perfusion Reduced diuretic responsiveness Worsening renal function	Nephroangio-sclerosis, chronic interstitial nephropathy	ADHF, severe arrythmias, shock	LVH, Dilatative myocardiopathy, CHF	ADHF – AKI
Criteria for sequelae	RIFLE R,I,F by creatinine or urine output.	CKD stage 1–5	Pulmonary edema, electrolyte imbalance arrythmias, cardiac arrest, reduced myocardiac contractility ADHF, AHF, pericarditis	CHF Decreased EF Increased CVP NYHF class 3–4	RIFLE R,I,F or RRT
Cardiac markers	Troponin, BNP, MPO	BNP, CRP	BNP, CRP	CRP	CRP, procalcitonin
Renal markers	Serum cystatine, creatinine, NGAL. urinary KIM-1, IL-18, NGAL, NAG	Creatinine, cystatic C, urea, uric acid, CRP Decreased GFR	Serum cystatine, creatinine, NGAL. urinary KIM-1, IL-18, NGAL, NAG	Creatinine, cystatic C, urea, uric acid CRP Decreased GFR	Creatinine, NGAL, IL-18, KIM-1, NAG

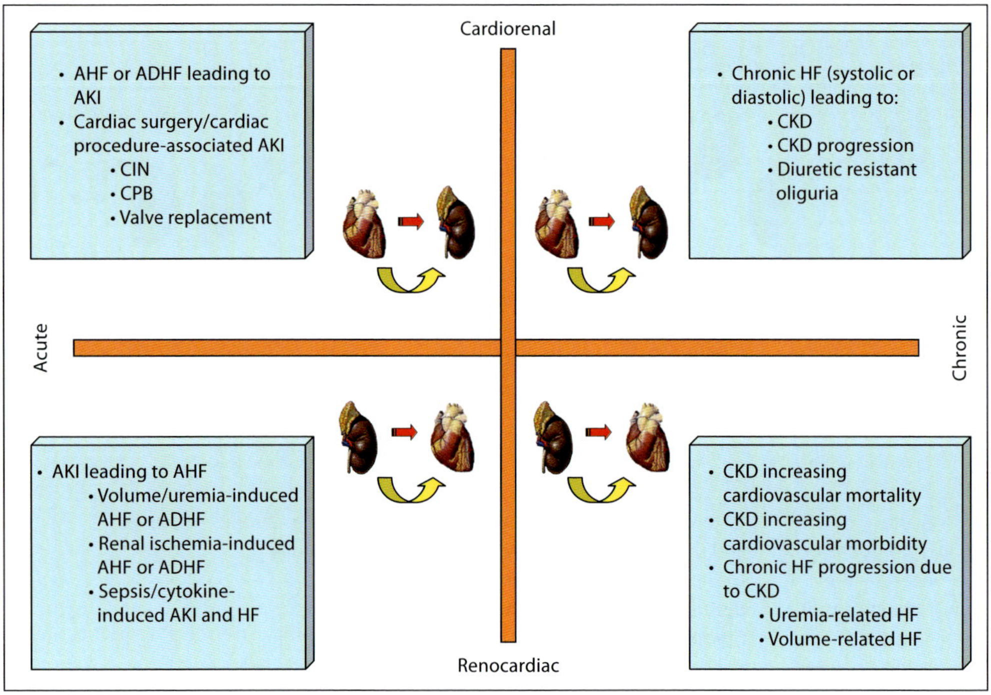

Fig. 1. CRS domain map. The timeframe and the bidirectionality of the different syndromes are described in the different subtypes (acute and chronic; cardiorenal and renocardiac). Every syndrome is then susceptible of transforming into another due to instauration of vicious circles. HF = Heart failure; CIN = contrast-induced nephropathy; CPB = cardiopulmonary bypass.

predisposes them to acute kidney injury (AKI), which has been associated with higher risk of prolonged/re-hospitalization, cardiovascular (CV) and all-cause mortality, and faster progression to chronic kidney disease (CKD) stages 4–5 [5–9]. The incidence estimates for AKI or worsening renal function (WRF) associated with AHF and acute coronary syndrome range between 24–45 and 9–19%, respectively [4]. The broad range in reported incidence is largely attributable to variations in the definitions of WRF as well as retrospective and/or post-hoc analysis data collection [5–9].

The mechanisms by which the onset of AHF or ADHF leads to AKI are multiple and complex [6]. They have been widely described in recent publications [3, 4]. The clinical importance of each of these mechanisms is likely to vary from patient to patient. The identified involved mechanisms are hemodynamic mechanisms in which a reduced cardiac output, together with an increased renal venous congestion may affect intrarenal circulation and glomerular filtration rate. Neurohormonal biofeedback mechanisms including renin-angiotensin-aldosterone system activation, a nonosmotic secretion of vasopressin and a parallel sympathetic nervous system-mediated vasoconstriction may lead to perpetuation and extension of the primary insult. We may also suggest an immune-mediated damage where the activation of monocytes and the direct effect of proinflammatory mediators may play an important role in renal tissue apoptosis and damage [10]. Finally, exogenous nephrotoxic agents such as contrast media, aminoglycosides, diuretics or angiotensin-converting enzyme inhibitors may represent another important source of kidney insult in patients with AHF or ADHF.

Humoral Signaling in CRS Type 1

The multiple factors involved in the development of AKI during heart failure describe a pathogenesis of AKI accounting for multiple pathways. Nevertheless, the occurrence of AKI even in patients with preserved or intact ejection fraction, the worsening of renal function even in the absence of fall in cardiac output, and the concept that AKI does not necessarily mean acute tubular necrosis have recently spurred new interest in humoral signaling and the inflammatory pathways to explain distant organ damage and, in particular, kidney injury in heart failure. It is known that ADHF implies an increase in circulating cytokines, but the level of this inflammatory response has not been quantified and it has not been determined in single patients.

We performed a study on 16 patients with ADHF analyzing renal function throughout hospital admission. Eight of the patients developed AKI according to the RIFLE criteria. In all patients we collected serum samples to analyze the possibility of different inflammatory patterns in patients with or without AKI development. Urea creatinine and brain natriuretic peptides (BNP) were not significantly different at admission in the two groups (A = AKI; B = non-AKI) and the amount of diuretics administered was similar. Serum samples collected at admission in all patients were incubated with human monocytic cultured cells (U-937) and the level of apoptosis was evaluated up to 96 h. Our hypothesis was that patients who develop AKI might have an inflammatory pattern due to various mediators that induce an early pathological apoptosis in normal monocytes. In this case, the humoral signal for a distant organ effect could be proved demonstrating an important contribution of the inflammatory pathway to the development of CRS type 1. In figure 2, cell apoptosis is reported in the two groups, demonstrating a significant difference at various incubation time points.

This is the first piece of evidence that CRS type 1 is an inflammatory disorder or at least the contribution of the inflammatory pathways in its pathogenesis is fundamental for the occurrence of distant organ damage.

Type 2 CRS – Chronic Cardiorenal Syndrome

Chronic heart failure patients with secondary WRF are associated with significantly increased adverse CV mortality and prolonged hospitalization. However, chronic heart disease and CKD frequently coexist in the same patient, and

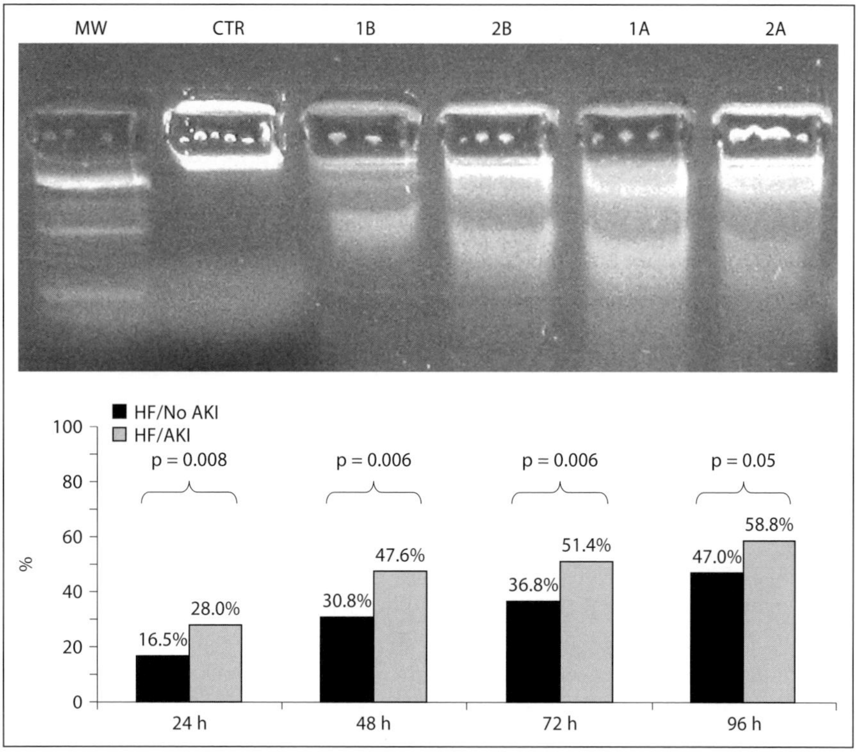

Fig. 2. On the electrophoretic gel it is possible to note a more intense pattern in patients in group A (CRS type 1) than in patients group B (ADHF without AKI). The difference in apoptosis between the two groups is significant at various time-points. HF = Heart failure; MW = marker; CTR = control.

often the clinical scenario does not permit the ability to distinguish which disease came first. Notably, there are about 45–63.6% of chronic heart failure patients who have evidence of CKD with an estimated glomerular filtration rate <60 ml/min/1.73 m^2 [11, 12]. The mechanisms causing secondary WRF in these patients likely differ from CRS type 1. Concomitant hemodynamic alterations with poor cardiac contractility and neurohormonal abnormalities with excessive production of vasoconstrictive mediators (epinephrine, angiotensin, endothelin), as well as altered sensitivity and/or release of endogenous vasodilatory factors (natriuretic peptides, nitric oxide), are commonly present in chronic heart failure patients. Hence, the kidneys that are chronically under a hypoperfusion state will be highly susceptible to any minor or major insults. Genetic and acquired micro- or macrovascular risk factors such as diabetic mellitus, hypertension and dyslipidemia may contribute to initiate and perpetuate the kidney damage. Moreover, subclinical inflammation, endothelial dysfunction and accelerated atherosclerosis determine a progressive sclerosis and fibrosis in the kidney tissue with a progressive loss of nephrons. These processes result in an initial phase of CKD that contributes to pathological conditions such as anemia, calcium-phosphate abnormalities, reduction of vitamin D receptor activation, hyperparathyroidism and hypertension. A vicious cycle is activated and a combined worsening of heart and kidney function inevitably occurs.

Type 3 CRS – Acute Renocardiac Syndrome

The development of AKI as a primary event and subsequently leading to cardiac injury and/or dysfunction has been recently observed with high prevalence. Type 3 CRS is associated with poor short- and long-term outcomes. Unfortunately, it has not been systematically evaluated or studied because of the heterogeneity in the etiologies of AKI, the variability in the definition of AKI and the failure of many clinical studies of AKI to report the occurrence of acute cardiac dysfunction as outcomes. AKI can affect the heart through several pathways. Fluid overload can contribute to the development of pulmonary edema. Electrolyte imbalances such as hyperkalemia can contribute to arrhythmias, which may cause cardiac arrest. Untreated uremia affects myocardial contractility through the accumulation of myocardial depressant factors, pulmonary vasoconstriction and/or pericarditis [13]. In addition, renal ischemia itself may induce an inflammatory-mediated cardiac injury via cytokine induction, leukocyte infiltration and apoptosis [14], and possibly a central role of the monocyte may be envisaged. If AKI is severe and renal replacement therapy is necessary, conventional hemodialysis with rapid fluid and electrolyte shifts can induce hypotension, arrhythmias and myocardial ischemia. Continuous renal replacement therapy, which minimizes such CV instability, appears to be physiologically safer and more logical in this setting.

Type 4 CRS – Chronic Renocardiac Syndrome

CKD is classified into 5 stages. Several studies have evaluated the CV outcomes in selected CKD-specific populations [15–18]. Increasing evidence show that CKD leads to myocardial interstitial fibrosis, ventricular hypertrophy, and systolic and diastolic dysfunction [17, 18]. CKD is an independent risk factor for CV outcomes in patients with and without chronic heart failure and post-cardiac surgery. Even moderate degrees of renal insufficiency are independently associated with an increased risk of death in patients with heart failure and left ventricular (LV) systolic dysfunction. It is well understood that patients with moderate to severe renal insufficiency not only carry a high burden of traditional CV risk factors, but are also exposed to uremia-specific risk factors that in concert induce an excessively increased CV mortality [18]. Uremia, acidosis, inflammation, malnutrition, endothelial dysfunction, anemia and calcium-phosphate abnormalities seem to play an important role in this setting. When patients are on dialysis, specific causes of CV events may arise from subclinical inflammation, chronic exposure with artificial biomaterials, dialysate impurity and infection at the vascular access site.

Type 5 CRS – Secondary CRS
There is limited data on the epidemiology of secondary CRS in general due to the large number of potential contributing acute and chronic systemic conditions that affect both the kidneys and heart. Although there is an appreciation that mortality increases in critical illness as more organs fail, there is limited insight into how combined renal and CV failure may differently affect such outcome compared to, for example, combined pulmonary and renal failure. Nonetheless, it is clear that several acute and chronic diseases can affect both organs simultaneously and that the disease induced in one can affect the other and vice versa. In the acute setting, severe sepsis represents the most common and serious condition which can affect both organs. It can induce AKI while leading to profound myocardial depression. The mechanisms responsible for such changes are poorly understood, but they may involve the effect of tumor necrosis factor and other humoral mediators on both organs [19, 20]. A different series of mechanisms are involved in chronic systemic disorders, such as amyloidosis or diabetes, where the combined damage is likely to be due to a severe endothelial dysfunction and accelerated atherosclerosis.

A description of the epidemiology and pathophysiology of heart-kidney interaction, stratified by the CRS subtypes, is a critical initial step towards understanding the overall burden of disease for each CRS subtype, their natural history, associated morbidity and mortality, potential prevention or treatment strategies, and health resource implications.

Biomarkers in Cardiorenal Syndrome

Recent studies have attempted to evaluate the utility of biomarkers for assessing CKD risks in CVD patients and CV risks in CKD patients [21–25]. There seems to be a crucial role of biomarkers in the early diagnosis of evolving heart failure or AKI in different subtypes of CRS. This may allow earlier changes in the management, such as stopping harmful interventions or to make early therapeutic interventions and protective/preventive strategies. Biomarkers could also help to make a more accurate differential diagnosis of heart failure and AKI, which would help prognostic stratification of the disorders, staging of diseases, and assessment of current and future severity of injury.

Several cardiac biomarkers like BNPs, troponins, myeloperoxidase, asymmetric dimethylarginine, plasminogen-activator inhibitor type I, homocysteine, C-reactive protein, CK-MB, myoglobin, serum amyloid A protein, ischemia-modified albumin and advanced glycation end-products have been demonstrated to correlate with CV outcomes in CKD patients [21–27]. It is well known that troponins correlate with the severity of cardiac injury, even in the absence of typical clinical features [23, 25]. BNPs are elevated in patients with CRS type 1 in which AKI occurs as a consequence of AHF or ADHF. Importantly, both troponins and BNPs have shown their prognostic utility in patients with various stages of renal insufficiency [22–24, 26–28]. Recently, Hayashi et al. [29] found that troponin T was a useful screening tool for asymptomatic coronary artery disease in CKD patients. Furthermore, elevated troponin T and NT-pro BNP levels were found to be correlated with hypervolemia and was able to identify a subgroup of end-stage renal disease (ESRD) patients with a 2- to 5-fold increase in mortality despite being asymptomatic [30]. In chronic peritoneal dialysis patients, NT-pro BNP was a marker of LV dysfunction and CV congestion, which was associated with high CV death and all-cause mortality [31].

Renal biomarkers like cystatin C and neutrophil gelatinase-associated lipocalin (NGAL) have recently been evaluated as diagnostic and prognostic markers for CV outcomes in CKD patients [32–36]. Higher levels of cystatin C have been shown to be directly involved in the atherosclerotic process [36] and is associated with increased LV mass and its concentricity independent of renal function [37]. Cystatin C appears to be a marker of CV risk at intermediate-term follow-up of acute coronary syndrome and ADHF patients [38, 39]. Similarly, increased NGAL expression has been demonstrated in atherosclerotic plaque and failing myocardium in coronary artery disease or heart failure patients [40, 41]. Its levels were significantly higher in the presence of these disorders and were correlated with the disease severity independent of coexisting renal injury [41, 42]. ADHF patients with an elevated admission serum NGAL had a higher risk of WRF [33]. In a small study of 46 elderly chronic heart failure patients, higher levels of plasma NGAL were found in parallel with the clinical severity of chronic heart failure, the highest levels being reached in NYHA class IV patients and were associated with higher mortality [35]. Several studies have demonstrated a significant rise in NGAL levels in AKI patients secondary to cardiac surgery, sepsis, ischemia and nephrotoxins [43].

Management of Cardiorenal Syndrome

Because the pathophysiology of CRS is complex, multimodal preventive and therapeutic strategies working via multiple targets are needed. To date, there are no guidelines for managing patients with AKI or CRS. Physicians should rely on the knowledge accumulated for the treatment of the specific organ diseases [44–46]. Hopefully, the new CRS classification will help to design the trials necessary to generate the missing evidence in this field.

In CRS type 1 and 2 patients, elimination of cause and/or disease leading to cardiac damage and heart failure progression, coronary artery disease risk factor modification, blood pressure control, compliance with dietary sodium restriction and drug treatments are crucial steps in their management. The main drugs consist of renin-angiotensin-aldosterone system blockers and β-adrenergic blockers to mitigate the overactivation of renin-angiotensin-aldosterone system and sympathetic system in these groups of patients unless contraindicated [44, 47–49]. Vasodilators and loop diuretics are widely used in cases of ADHF [50]. However, these drugs do carry some risks in WRF. Other drugs like vasopressin receptor-2 antagonists, phosphodiesterase inhibitors (milrinone) and lusitropic agents (levosimendan) have been used in some cases, but without any clear survival benefit. Extracorporeal ultrafiltration may be useful in AHF associated with diuretic resistance, resulting in significant fluid removal and weight loss. If systemic hypotension persists, inotropic agents may then be considered, along with elective ventilation and/or intra-aortic balloon pumping. Depending on pre-existing comorbidities and the underlying etiology, LV assist devices may be needed as a bridge to transplantation or cardiac surgery. Cardiac resynchronization therapy is now recommended for symptomatic chronic heart failure patients (NYHA III-IV) with poor LV ejection fraction and QRS prolongation [44] and implantable cardiac defibrillators for both survivors of cardiac arrest and/or sustained ventricular arrhythmias as well as for symptomatic chronic heart failure patients with impaired LV ejection fraction.

In CRS type 3 and 4 patients, the clinical problem is often sodium and water retention. Prompt and adequate treatment of hypervolemia may help prevent cardiac decompensation. Uremic abnormalities and hyperkalemia may further contribute to secondary heart dysfunction and therefore all efforts should be made to avoid advanced disorders. Mediators of inflammation in renal ischemia [14] have also been suspected of playing an important role in negative organ crosstalk with the heart. Strategies directed towards lessening these mechanisms should be attempted to reduce the combination of multiple factors in the pathogenesis of heart dysfunction. Moreover, it is important to retard the progression of CKD, which may result in cardioprotective strategies. Early treatment of anemia, minimizing vascular calcification and vitamin D receptor abnormalities may help to prevent progression of CKD and may help limit CV complications. Special measures such as biocompatible materials, convective therapy and ultrapure dialysate can also be implemented to reduce the incidence of CV complications in hemodialysis patients lead to reduced rates of CRS type 4.

Finally, in CRS type 5 patients, treatment of the primary illness (diabetes mellitus, amyloidosis, sepsis, rhabdomyolysis, hemorrhagic shock, etc.) generally improves both heart and kidney function.

Conclusions

The new definition and classification system for CRS seems to represent an important advance and spur new interest in collaborative nephrologicalcardiological efforts. The complexity of CRS and the specificity of care necessary to offer the best therapy to these patients demand a multidisciplinary approach which combines the expertise of cardiology, nephrology and critical care medicine. This multidisciplinary effort should facilitate a better appreciation of the interaction between the heart and kidneys during dysfunction of each or both organs with significant practical clinical implications. Epidemiological studies in different settings could be carried out classifying patients according to specific disorders or CRS subtypes. Investigators can identify aspects of each single syndrome subtype that carry clear priorities for further research. It is true that patients may move from one subtype to another, but this should not discourage a categorization of a patient in a specific timeframe of his medical history. It may be difficult to distinguish between type 2 and type 4 in some cases, but at least a careful discussion about prevention and therapeutic priorities should be done collegially between nephrologists and cardiologists. Hopefully, all these efforts will contribute to an increased awareness and physiological understanding leading to some innovative therapeutic approaches and possibly to improved clinical outcomes.

References

1 Ronco C: Cardiorenal and renocardiac syndromes: clinical disorders in search of a systematic definition. Int J Artif Organs 2008;31:1–2.

2 Ronco C, House AA, Haapio M: Cardiorenal syndrome: refining the definition of a complex symbiosis gone wrong. Intensive Care Med 2008;34:957–962.

3 Ronco C, Haapio M, House AA, et al: Cardiorenal syndrome. J Am Coll Cardiol 2008;52:1527–1539.

4 Ronco C, McCullough P, Anker SD, et al: Cardio-renal syndromes: report from the consensus conference of the acute dialysis quality initiative. Eur Heart J 2010;31:703–711.

5 Adams KF Jr, Fonarow GC, Emerman CL, et al: Characteristics and outcomes of patients hospitalized for heart failure in the United States: rationale, design, and preliminary observations from the first 100,000 cases in the Acute Decompensated Heart Failure National Registry (ADHERE). Am Heart J 2005;149:209–216.

6 Liang KV, Williams AW, Greene EL, et al: Acute decompensated heart failure and the cardiorenal syndrome. Crit Care Med 2008;36(suppl 1):S75–S88.

7 Gottlieb SS, Abraham W, Butler J, et al: The prognostic importance of different definitions of worsening renal function in congestive heart failure. J Card Fail 2002;8:136–141.

8 Newsome BB, Warnock DG, McClellan WM, et al: Long-term risk of mortality and end-stage renal disease among the elderly after small increases in serum creatinine level during hospitalization for acute myocardial infarction. Arch Intern Med 2008;168:609–616.

9 Goldberg A, Hammerman H, Petcherski S, et al: Inhospital and 1-year mortality of patients who develop worsening renal function following acute ST-elevation myocardial infarction. Am Heart J 2005;150:330–337.

10 Cao Z, Cooper ME: Role of angiotensin II in tubulointerstitial injury. Semin Nephrol 2001;21:554–562.

11 Hillege HL, Nitsch D, Pfeffer MA, et al: Renal function as a predictor of outcome in a broad spectrum of patients with heart failure. Circulation 2006;113:671–678.

12 Heywood JT, Fonarow GC, Costanzo MR, et al: High prevalence of renal dysfunction and its impact on outcome in 118,465 patients hospitalized with acute decompensated heart failure: a report from the ADHERE database. J Card Fail 2007;13:422–430.
13 Meyer TW, Hostetter TH: Uremia. N Engl J Med 2007;357:1316–1325.
14 Kelly KJ: Acute renal failure: much more than a kidney disease. Semin Nephrol 2006;26:105–113.
15 McCullough PA, Jurkovitz CT, Pergola PE, et al: Independent components of chronic kidney disease as a cardiovascular risk state: results from the Kidney Early Evaluation Program (KEEP). Arch Intern Med 2007;167:1122–1129.
16 Go AS, Chertow GM, Fan D, et al: Chronic kidney disease and the risks of death, cardiovascular events, and hospitalization. N Engl J Med 2004;351:1296–1305.
17 Gross ML, Ritz E: Hypertrophy and fibrosis in the cardiomyopathy of uremia—beyond coronary heart disease. Semin Dial 2008;21:308–318.
18 Remppis A, Ritz E: Cardiac problems in the dialysis patient: beyond coronary disease. Semin Dial 2008;21:319–325.
19 Cunningham PN, Dyanov HM, Park P, et al: Acute renal failure in endotoxemia is caused by TNF acting directly on TNF receptor-1 in kidney. J Immunol 2002;168:5817–5823.
20 Kumar A, Paladugu B, Mensing J, et al: Nitric oxide-dependent and -independent mechanisms are involved in TNF-alpha -induced depression of cardiac myocyte contractility. Am J Physiol Regul Integr Comp Physiol 2007;292:R1900–R1906.
21 Parikh SV, de Lemos JA: Biomarkers in cardiovascular disease: integrating pathophysiology into clinical practice. Am J Med Sci 2006;332:186–197.
22 Carr SJ, Bavanandan S, Fentum B, et al: Prognostic potential of brain natriuretic peptide (BNP) in predialysis chronic kidney disease patients. Clin Sci (Lond) 2005;109:75–82.
23 Howie-Esquivel J, White M: Biomarkers in acute cardiovascular disease. J Cardiovasc Nurs 2008;23:124–131.
24 Needham DM, Shufelt KA, Tomlinson G, et al: Troponin I and T levels in renal failure patients without acute coronary syndrome: a systematic review of the literature. Can J Cardiol 2004;20:1212–1218.
25 Cameron SJ, Sokoll LJ, Laterza OF, et al: A multi-marker approach for the prediction of adverse events in patients with acute coronary syndromes. Clin Chim Acta 2007;376:168–173.
26 Sommerer C, Beimler J, Schwenger V, et al: Cardiac biomarkers and survival in haemodialysis patients. Eur J Clin Invest 2007;37:350–356.
27 Austin WJ, Bhalla V, Hernandez-Arce I, et al: Correlation and prognostic utility of B-type natriuretic peptide and its amino-terminal fragment in patients with chronic kidney disease. Am J Clin Pathol 2006;126:506–512.
28 Hickman PE, McGill DA, Talaulikar G, et al: Prognostic efficacy of cardiac biomarkers for mortality in dialysis patients. Intern Med J 2009;39:812–818.
29 Hayashi T, Obi Y, Kimura T, et al: Cardiac troponin T predicts occult coronary artery stenosis in patients with chronic kidney disease at the start of renal replacement therapy. Nephrol Dial Transplant 2008;23:2936–2942.
30 Codognotto M, Piccoli A, Zaninotto M, et al: Effect of a dialysis session on the prognostic values of NT-proBNP, troponins, endothelial damage and inflammation biomarkers. J Nephrol 2010;23:465–471.
31 Wang AY, Lam CW, Yu CM, et al: N-terminal pro-brain natriuretic peptide: an independent risk predictor of cardiovascular congestion, mortality, and adverse cardiovascular outcomes in chronic peritoneal dialysis patients. J Am Soc Nephrol 2007;18:321–330.
32 Iwanaga Y, Miyazaki S: Heart failure, chronic kidney disease, and biomarkers—an integrated viewpoint. Circ J 2010;74:1274–1282.
33 Aghel A, Shrestha K, Mullens W, et al: Serum neutrophil gelatinase-associated lipocalin (NGAL) in predicting worsening renal function in acute decompensated heart failure. J Card Fail 2010;16:49–54.
34 Poniatowski B, Malyszko J, Bachorzewska-Gajewska H, et al: Serum neutrophil gelatinase-associated lipocalin as a marker of renal function in patients with chronic heart failure and coronary artery disease. Kidney Blood Press Res 2009;32:77–80.

35 Bolignano D, Basile G, Parisi P, et al: Increased plasma neutrophil gelatinase-associated lipocalin levels predict mortality in elderly patients with chronic heart failure. Rejuvenation Res 2009;12:7–14.
36 Ix JH, Shlipak MG, Chertow GM, et al: Association of cystatin C with mortality, cardiovascular events, and incident heart failure among persons with coronary heart disease: data from the Heart and Soul Study. Circulation 2007;115:173–179.
37 Patel PC, Ayers CR, Murphy SA, et al: Association of cystatin C with left ventricular structure and function: the Dallas Heart Study. Circ Heart Fail 2009;2:98–104.
38 Taglieri N, Fernandez-Berges DJ, Koenig W, et al: Plasma cystatin C for prediction of 1-year cardiac events in Mediterranean patients with non-ST elevation acute coronary syndrome. Atherosclerosis 2010;209:300–305.
39 Campbell CY, Clarke W, Park H, et al: Usefulness of cystatin C and prognosis following admission for acute heart failure. Am J Cardiol 2009;104:389–392.
40 Hemdahl AL, Gabrielsen A, Zhu C, et al: Expression of neutrophil gelatinase-associated lipocalin in atherosclerosis and myocardial infarction. Arterioscler Thromb Vasc Biol 2006;26:136–142.
41 Yndestad A, Landro L, Ueland T, et al: Increased systemic and myocardial expression of neutrophil gelatinase-associated lipocalin in clinical and experimental heart failure. Eur Heart J 2009;30:1229–1236.
42 Zografos T, Haliassos A, Korovesis S, et al: Association of neutrophil gelatinase-associated lipocalin with the severity of coronary artery disease. Am J Cardiol 2009;104:917–920.
43 Mori K, Nakao K: Neutrophil gelatinase-associated lipocalin as the real-time indicator of active kidney damage. Kidney Int 2007;71:967–970.
44 Dickstein K, Cohen-Solal A, Filippatos G, et al: ESC guidelines for the diagnosis and treatment of acute and chronic heart failure 2008: the Task Force for the Diagnosis and Treatment of Acute and Chronic Heart Failure 2008 of the European Society of Cardiology. Developed in collaboration with the Heart Failure Association of the ESC (HFA) and endorsed by the European Society of Intensive Care Medicine (ESICM). Eur J Heart Fail 2008;10:933–89.
45 K/DOQI clinical practice guidelines for chronic kidney disease: evaluation, classification, and stratification. Am J Kidney Dis 2002;39(2 suppl 1):S1–S266.
46 Mactier RA, Tomson CR: Executive summary of the fifth edition of the Renal Association Clinical Practice Guidelines (2009–2012). Nephron Clin Pract 2011;118 (suppl 1):5–11.
47 Bangalore S, Messerli FH, Kostis JB, et al: Cardiovascular protection using beta-blockers: a critical review of the evidence. J Am Coll Cardiol 2007;50:563–572.
48 Pfeffer MA, Swedberg K, Granger CB, et al: Effects of candesartan on mortality and morbidity in patients with chronic heart failure: the CHARM-Overall programme. Lancet 2003;362:759–766.
49 Pitt B, Zannad F, Remme WJ, et al: The effect of spironolactone on morbidity and mortality in patients with severe heart failure. Randomized Aldactone Evaluation Study Investigators. N Engl J Med 1999;341:709–717.
50 Hollenberg SM: Vasodilators in acute heart failure. Heart Fail Rev 2007;12:143–147.

Claudio Ronco, MD
Department of Nephrology, Dialysis & Transplantation
International Renal Research Institute Vicenza
S. Bortolo Hospital, Viale Rodolfi, 37
Vicenza, IT–36100 (Italy)
Tel. +39 44 475 3650, E-Mail cronco@goldnet.it

and HRS underwent LT because most patients died before LT as a consequence of the rapid evolution of type 1 HRS. Thus, new treatment options were developed for HRS at the end of 1990s. Among these options, the use of albumin and vasoconstrictors proved to be an effective 'treatment bridge' towards LT, increasing the number of patients with type 1 HRS who could undergo LT and ameliorating their LT outcome [16].

Vasoconstrictors for the Treatment of Type 1 Hepatorenal Syndrome

The administration of vasoconstrictors and albumin in patients with type 1 HRS is based on the current knowledge of the pathophysiology of type 1 HRS. Severe renal arterial vasoconstriction, which is the extreme renal functional abnormality that can occur in patients with cirrhosis and ascites, represents the pathophysiological basis of HRS [2]. As previously discussed, it develops in the context of a marked reduction of effective circulating volume, which is related to a splanchnic arterial vasodilation and inadequate cardiac output (fig. 1) and which implies an extreme overactivation of the endogenous systemic vasoconstrictor systems, i.e. the renin-angiotensin system, sympathetic nervous system and nonosmotic release of vasopressin. In using this method, the final aim of this therapeutical approach is to reduce severe renal arterial vasoconstriction [17]. In small prospective pilot studies as well as in retrospective studies, it has been demonstrated that the prolonged use of a vasoconstrictor derived from vasopressin (ornipressin [18, 19] or terlipressin [20–31]) or of an α-agonist vasoconstrictor (midodrine plus octreotide, noradrenaline alone) [32–37] in association with human albumin is capable of recovering renal function in patients with type 1 HRS. These studies have shown that a vasoconstrictor plus albumin can recover renal function in 40–60% of cases. Recurrence of HRS after treatment withdrawal (a sharp increase in serum creatinine within a few days) occurs in approximately 20% of patients. Treatment with terlipressin should be repeated in these patients and this measure is frequently effective. This can result in long-term treatment in some cases [38].

Among vasoconstrictors, to this day terlipressin is the most widely used in the treatment of type 1 HRS [5, 20–31]. Terlipressin has been used in more than 200 patients either as an intravenous bolus starting at an initial dose of 0.5 mg every 4–6 h [20–29], or continuous intravenous infusion starting at an initial dose of 2 mg/day [30, 31]. In patients without a response (no significant reduction of serum creatinine within 3 days), the initial dose of terlipressin is doubled. The maximal doses of terlipressin used in the treatment of type 1 HRS are 2 mg every 4–6 h by intravenous boluses, or 12 mg/day by continuous intravenous infusion, respectively. Complete reversal (defined by a decrease of serum creatinine with a final value <1.5 mg/dl) or partial reversal (defined with a decrease of serum creatinine >50% with a final value ≥1.5 mg/dl) of type 1 HRS has been observed in almost 59% of patients [5].

In most studies, terlipressin has been used together with albumin starting with a priming dose of 1 g/kg of body weight followed by 20–40 g/day, monitoring central venous pressure. In two studies in which terlipressin was also given alone [25, 28], reversal of renal failure was lower than in the studies in which terlipressin was associated with albumin. Up to now, only two randomized controlled clinical trials comparing terlipressin and albumin with albumin alone have been published [39, 40]. Both trials confirmed the effectiveness of terlipressin in recovering renal function in patients with type 1 HRS. Nevertheless, they failed in improving survival in patients with type 1 HRS. Nonetheless, a systematic review of all the randomized controlled clinical trials has recently shown that terlipressin plus albumin may prolong only the 15-day survival in patients with type 1 HRS [41]. This evidence enhances the meaning of the use of terlipressin and albumin as a bridge treatment towards LT. Nonetheless, it appears more and more evident that this treatment in clinical practice is often used also in patients with type 1 HRS who are not candidates for LT [5].

The small effect of terlipressin and albumin on survival may be explained by taking into account that the prognosis in these patients is largely dependent upon only the degree of liver failure. A marked impairment of liver function represents a poor predictor not only for the response to treatment with terlipressin and albumin, but also for their overall survival [25, 26, 28]. In addition, terlipressin is targeted on splanchnic arterial vasodilation, but has no effect on the impaired cardiac output in these patients. This has been shown to play an important role in the pathophysiology of type 1 HRS. Thus, the only effect on cardiac output of this therapeutic approach up to now is associated with albumin infusion.

As far as nonresponders to terlipressin and albumin are concerned, it should be mentioned that it has only been recently shown that the predictive factors of nonresponse are not only related to the degree of impairment of liver function, i.e. a high baseline value of total serum bilirubin, but also to a milder hemodynamic response to the treatment [42] and an overall high baseline value of serum creatinine. In particular, it has been shown that patients with a baseline serum creatinine lower than 3.0 mg/dl have the highest probability of responding to treatment [43]. This observation has relevant implications for the management of patients with type 1 HRS. Firstly, it is necessary to make the diagnosis as quickly as possible so as to begin treatment as early as possible. Secondly, it would most likely be appropriate to revise the limit of 2.5 mg/dl serum creatinine to make the diagnosis of HRS.

To conclude the discussion on nonresponders, two other aspects should be briefly considered: (1) the complexity of the nature of HRS, and (2) the complexity of its pathophysiology. It has been stated that HRS is a functional renal failure, but there is increasing evidence of the presence of renal vascular and tubular lesions in patients who developed this complication [44]. In addition, the aim of terlipressin and albumin is to improve renal perfusion through improvement of the effective circulating volume. Nevertheless, looking in experimental models of renal failure precipitated by bacterial infections as occurs in most patients who develop HRS, it is easy to see that many other factors are involved in determining the loss of glomerular filtration rate beyond the reduction of renal blood flow [45, 46].

Vasocontrictors for the Treatment of Type 2 Hepatorenal Syndrome
Type 2 HRS is a more stable impairment of renal function in patients with cirrhosis and ascites. Thus, patients with type 2 HRS do not present with acute renal failure, but rather refractory ascites. Consequently, the effects of the vasoconstrictors and albumin in type 2 HRS treatment have not been largely investigated. Nonrandomized studies which were performed on a small number of patients with type 2 HRS, however, have shown that the percentage of response to treatment in terms of recovered renal function does not seem to be different from that observed in patients with type 1 HRS [25, 26, 47], while survival appears longer (100% at 3 months).

Conclusions

The use of terlipressin and albumin in the treatment of type 1 HRS represents a milestone in the treatment of complications in patients with advanced cirrhosis. Its rationale is closely related to our current knowledge on the pathophysiology of type 1 HRS. The use of terlipressin and albumin has proven to be effective in recovering renal function and has increased short-term survival in responders, making it possible to increase the number of patients who undergo orthotopic LT. In addition, it has changed the outcome of LT in these patients since it has reduced the post-LT need of RRT and increased the post-LT survival. Despite these encouraging results, it should be noted that a recovery of renal function can be achieved in less than 50% of patients with type 1 HRS and, that the recovery of renal function may be partial in patients who are defined as full responders. Terlipressin plus albumin is a therapeutical option with some inherent limitations related to the complexity of both, the nature and the pathophysiology of HRS, and extrinsic limitations related to the schedule which has frequently been used up to now, i.e. the administration of terlipressin by intravenous boluses. Thus, new therapeutic options that explore the complexity of the nature and pathophysiology of HRS should be investigated. In the meantime, the use of vasoconstrictors and albumin should be optimized. Therefore, a very dense and ambitious research agenda on HRS will await hepatologists in the near future (fig. 2).

- Improving the diagnosis of HRS with new biomarkers
- Improving our understanding of the pathophysiology of HRS
- Starting treatment with terlipressin and albumin early
- Improving the hemodynamic monitoring of treatment
- Targeting a ΔMAP during treatment
- Testing the association of terlipressin and albumin with:
  - An adequate inotropic agent
  - An antioxidant agent
  - A TNF-α modulatory agent
- Testing the association of terlipressin and albumin with non-pharmacological options

Fig. 2. Research agenda in HRS. MAP = Mean arterial pressure.

References

1 Gines A, Escorsell A, Gines P, et al: Incidence, predictive factors and prognosis of the hepatorenal syndrome in cirrhosis with ascites. Gastroenterology 1993;105:229–236.

2 Schrier RW, Arroyo V, Bernardi M, et al: Peripheral arteriolar vasodilation hypothesis: a proposal for the initiation of renal sodium and water retention in cirrhosis. Hepatology 1988;8:1151–1157.

3 Arroyo V, Gines P, Gerbes AL, et al: Definition and diagnostic criteria of refractory ascites and hepatorenal syndrome in cirrhosis. International Ascites Club. Hepatology 1996;23:164–176.

4 Salerno F, Gerbes A, Gines P, Wong F, Arroyo V: Diagnosis, prevention and treatment of the hepatorenal syndrome in cirrhosis. A consensus workshop of the International Ascites Club. Gut 2007;56:1310–1318.

5 Gines P, Guevara M, Arroyo V, Rodes J: Hepatorenal sindrome. Lancet 2003;362:1819–1827.

6 Follo A, Llovet JM, Navasa M, et al: Renal impairment after spontaneous bacterial peritonitis in cirrhosis: incidence, clinical course, predictive factors and prognosis. Hepatology 1994;20:1495–1501.

7 Fernandez J, Navasa M, Planas R, et al: Primary prophylaxis of spontaneous bacterial peritonitis delays hepatorenal syndrome and improves survival in cirrhosis. Gastroenterology 2007;133:818–824.

8 Fasolato S, Angeli P, Dallagnese L, et al: Renal failure and bacterial infections in patients with cirrhosis: epidemiology and clinical features. Hepatology 2007;45:223–229.

9 Terra C, Guevara M, Torre A, et al: Renal failure in patients with cirrhosis and sepsis unrelated to spontaneous bacterial peritonitis: value of MELD score. Gastroenterology 2005;129:1944–1953.

10 Ruiz del Arbor L, Urman J, Fernandez J, et al: Systemic, renal and hepatic haemodynamic derangement in cirrhotic patients with spontaneous bacterial peritonitis. Hepatology 2003;38:1210–1218.

11 Ruiz-del-Arbol L, Monescillo A, Arocena C, et al: Circulatory function and hepatorenal syndrome in cirrhosis. Hepatology 2005;42:439–447.

12 Alqahtani SA, Fouad TR, Lee SS: Cirrhotic cardiomyopathy. Sem Liver Dis 2008;28:59–69.

13 Gonwa TA, Klintmalm GB, Levy M, et al: Impact of pretransplant renal function on survival after liver transplantation. Transplantion 1995;59:361–365.

14 Navasa M, Feu F, Garcia-Pagan JC, Jimenez W, et al: Hemodynamic and humoral changes after liver-transplantation in patients with cirrhosis. Hepatology 1993;17:355–360.

15 Schlitt HJ, Barkmann A, Boker KH, et al: Replacement of calcineurin inhibitors with mycophenolate mofetil in liver-transplant patients with renal dysfunction: a randomised controlled study. Lancet 2001;357:587–591.
16 Restuccia T, Ortega R, Guevara M, et al: Effects of treatment of hepatorenal syndrome before transplantation on posttransplantation outcome. A case control study. Liver Transpl 2004;40:140–146.
17 Narahara Y, Kanazawa H, Taki Y, et al: Effects of terlipressin on systemic, hepatic and renal hemodynamics in patients with cirrhosis. J Gastroenterol Hepatol 2009;24:1791–1797.
18 Guevara M, Gines P, Fernandez-Esparrach G, et al: Reversibility of hepatorenal syndrome by prolonged administration of ornipressin and plasma volume expansion. Hepatology 1998;27:35241.
19 Gülberg V, Bilzer M, Gerbes AL: Long-term therapy and retreatment of hepatorenal syndrome type 1 with ornipressin and albumin. Hepatology 1999;30:870–875.
20 Ganne-Carrie N, Hadegue A, Mathurin P, et al: Hepatorenal syndrome: long-term treatment with terlipressin as bridge to liver transplantation. Dig Dis Sci 1996;41:1054–1056.
21 Le Moine O, El Nawar A, Jagodzinski R, et al: Treatment with terlipressin as a bridge to translantation in a patient with hepatorenal syndrome. Acta Gastroenterol Belg 1998;61;268–270.
22 Duhamel C, Mauillon J, Berkelmans J, Bourienne A, Tranvuez JL: Hepatorenal syndrome in cirrhotic patients: terlipressin is a safe and efficient treatment; propoanolol and digitalic treatment: precipitating and preventing factors? Am J Gastroenterol 2000;95:2984–2985.
23 Halimi C, Bonnard P, Bernard B, et al: Effect of terlipressin (Glypressin) on hepatorenal syndrome in cirrhotic patients: results of a multicentre pilot study. Eur J Gastroenterol Hepatol 2002;14:153–158.
24 Mulkay JP, Louis H, Donckier V, et al: Long-term terlipressin administration improves renal function in cirrhotic patients with type 1 hepatorenal syndrome: a pilot study. Acta Gastroelerologica Belg 2001;64:15–19.
25 Uriz J, Gines P, Cardenas A, et al: Terlipressin plus albumin infusion: an effective and safe therapy of hepatorenal syndrome. J Hepatol 2000;33:43–48.
26 Ortega R, Ginès P, Uriz J, et al: Terlipressin therapy with and without albumin for patients with hepatorenal syndrome: results of a prospective, nonrandomized study. Hepatology 2002;36:941–948.
27 Solanki P, Chawala A, Garg R, et al: Beneficial effects of terlipressin in hepatorenal sindrome: a prospective, randomized placebo-controlled clinical trial. J Gastroenterol Hepatol 2003;18:152–156.
28 Moreau R, Durand F, Poynard T, et al: Terlipressin in patients with cirrhosis and type 1 hepatorenal syndrome: a retrospective multicenter study. Gastroenterology 2002;122:923–930.
29 Colle I, Durand F, Pessione F, et al: Clinical course, predictive factors and prognosis in patients with cirrhosis and type 1 hepatorenal syndrome treated with terlipressin: a retrospective analysis. J Gastroenterol Hepatol 2002;17:882–888.
30 Angeli P, Guarda S, Fasolato S, et al: Switch therapy with ciprofloxacin versus intravenous ceftazidime in the treatment of spontaneous bacterial peritonitis in patients with cirrhosis: similar efficacy at lower cost. Aliment Pharmacol Ther 2006;23:75–84.
31 Gerbes AL, Huber E, Gülberg V: Terlipressin for hepatorenal syndrome: continuous infusion as an alternative to i.v. bolus administration. Gastroenterology 2009;137:1179.
32 Angeli P, Volpin R, Gerunda G, et al: Reversal of type 1 hepatorenal syndrome (HRS) with the combined administration of midodrine and octreotide. Hepatology 1999;29:1690–1697.
33 Wong F, Pantea L, Sniderman K: Midodrine, octreotide, albumin, and TIPS in selected patients with cirrhosis and type 1 hepatorenal syndrome. Hepatology 2004;40:55–64.
34 Esrailian E, Pantageo ER, Kyulo NL, Hu KQ, Runyon BA: Octretide/midodrine therapy significantly improves renal function and 30-day survival in patients with type 1 hepatorenal syndrome. Dig Dis Sci 2007;52:742–748.

35 Skagen C, Einstein M, Lucey MR, Said A: Combination treatment with octreotide, midodrine, and albumin improves survival in patients with type 1 and type 2 hepatorenal syndrome. J Clin Gastroenterol 2009;43:680–685.
36 Devoux C, Zanditenas D, Hezode C, et al: Effects of noradrenalin and albumin in patients with type 1 hepatorenal syndrome: a pilot study. Hepatology 2002;36:374–380.
37 Sharma P, Kumar A, Sharma BC, Sarin SK: An open label, pilot, randomized controlled trial of noradrenaline versus terlipressin in the treatment of type 1 hepatorenal syndrome and predictors of response. Am J Gastroenterol 2008;103:1689–1697.
38 Piano S, Morando F, Fasolato S, et al: Continuous recurrence of type 1 hepatorenal syndrome and long-term treatment with terlipressin and albumin: a new exception to meld score in the allocation system to liver transplantation? J Hepatol 2011, E-pub ahead of print.
39 Sanyal A, Boyer T, Garcia-Tsao G, et al: A randomized, prospective, double blind placebo-controlled trial of terlipressin for type 1 hepatorenal syndrome. Gastroenterology 2008;134:1360–1368.
40 Martin-Llahi MM, Pepin MN, Guevara M, et al: Terlipressin and albumin versus albumin in patients with cirrhosis and hepatorenal syndrome: a randomized study. Gastroenterology 2008;134:1352–1359.
41 Gluud LL, Christensen K, Krag A: Systematic review of randomized trials on vasoconstrictor drugs for hepatorenal syndrome. Hepatology 2010;51:576–584.
42 Nazar A, Pereira GH, Guevara M, et al: Predictors of response to therapy with terlipressin and albumin in patients with cirrhosis and type 1 HRS Hepatology 2010;51:219–226.
43 Boyer TD, Sanyal AJ, Garcia-Tsao G, et al: Predictors of response to terlipressin plus albumin in hepatorenal syndrome (HRS) type 1: relationship of serum creatinine to hemodynamics. J Hepatol 2010 E-pub ahead of print.
44 Trawale JM, Paradis V, Rautou PE, et al: The spectrum of renal lesions in patients with cirrhosis: a clinicopathological study. Liver Int 2010;30:725–732.
45 Wang W, Mitra A, Poole B, et al: Endothelial nitric oxide synthase-deficient mice exhibit increased susceptibility to endotoxin-induced acute renal failure. Am J Physiol Renal Physiol 2004;287:F1044–F1048.
46 Zhou Z, Gengaro P, Wang W, et al: Role of NF-κB and PI 3-kinase/Akt in TNF-α-induced cytotoxicity in microvascular endothelial cells. Am J Physiol Renal Physiol 2008;295:F932–F941.
47 Alessandria C, Venon WD, Marzano A, Barletti C, Fadda M, Rizzetto M: Renal failure in cirrhotic patients: role of terlipressin in clinical approach to hepatorenal syndrome type 2. Eur J Gastroenterol Hepatol 2002;14:1363–1368.

Paolo Angeli, MD, PhD
Department of Clinical and Experimental Medicine
University of Padova
Via Giustiniani 2, cap. 35100, Padova (Italy)
E-Mail pangeli@unipd.it

Kellum JA, Ronco C, Vincent J-L (eds): Controversies in Acute Kidney Injury.
Contrib Nephrol. Basel, Karger, 2011, vol 174, pp 56–64

Clinical Consequences of Acute Kidney Injury

Eric A.J. Hoste[a,b] · Wouter De Corte[a,c]

[a]ICU Ghent University Hospital, Ghent University, Ghent, [b]Research Foundation-Flanders (FWO), and [c]Department of Anaesthesia and Intensive Care Medicine, Groeninge Hospital, Kortrijk, Belgium

Abstract

Acute kidney injury (AKI) can no longer be considered a surrogate marker for severity of illness. Recent epidemiologic data demonstrate the association of AKI and mortality. Even small decreases of kidney function are associated with increased mortality. Several clinical consequences of AKI may explain the association of AKI and mortality. Decreased free water clearance leading to volume overload contributes to morbidity and mortality, but also to deterioration of kidney function. Acid-base disorders and electrolyte abnormalities interfere with normal functioning of many processes in the body. Critically ill patients have an increased prevalence of infection. Infection and antimicrobial therapy can be the cause of AKI, but infection can also be a consequence of AKI. Finally, inadequate antimicrobial dosing probably plays an important role in the morbidity and mortality of AKI. These findings have led to a paradigm shift: patients die *because of* AKI rather than *with* AKI.

In the past, acute kidney injury (AKI) was considered a surrogate marker for severity of illness, and patient mortality was considered a consequence of the underlying disease [1]. Especially in intensive care unit (ICU) patients, AKI develops as a consequence of another disease, e.g. sepsis, cardiogenic shock or trauma, which in a certain number of patients will lead to mortality. However, there is an abundance of epidemiologic data that demonstrates that AKI itself also contributes to higher mortality. This is the case for the most severe form of AKI, where patients are treated with renal replacement therapy (RRT) [2–6]. Small decreases of kidney function are also associated with increased short-term mortality (hospital and 28-day). This has been demonstrated in various settings such as contrast-induced AKI, in patients who underwent cardiac surgery,

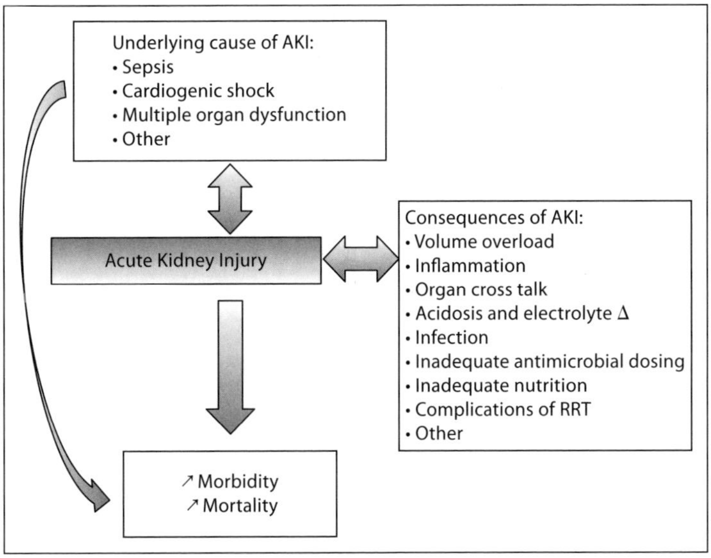

Fig. 1. Patients are dying of AKI and not with AKI. The interaction between the etiology of AKI and the consequences of AKI, and the impact on morbidity and mortality.

in hospitalized patients and in ICU patients [7–12]. When AKI is classified according to the newly developed sensitive RIFLE or AKI classification, all studies have demonstrated an association with hospital mortality [5, 13–15]. Other outcomes, such as length of hospital stay, readmission rate, development of end-stage kidney disease and long-term (1–10 years) mortality, are also affected by severe and less severe episodes of AKI during ICU stay [16–21].

In this overview we will discuss the clinical consequences of AKI that may explain the association of AKI and mortality (fig. 1).

Morbidity Caused by Acute Kidney Injury

Fluid Overload

Fluid resuscitation is one of the cornerstones for treatment of ICU patients with an episode of oliguria or developing AKI. The majority of AKI patients, especially those with severe AKI, will have decreased (free) water clearance. This will lead to accumulation of water, and several observational studies found that this is associated with worse outcome [22–24]. It is difficult to delineate whether fluid overload is only a surrogate marker of severity of illness or if it is in itself the cause of increased morbidity and mortality. Some arguments are in favor of the latter. Fluid overload may lead to a series of minor and major complications that may influence outcome. It may result in a broad range of complications such as development of tissue edema, ascites and eventually intra-abdominal hypertension and abdominal compartment syndrome, pleural effusion, and pulmonary edema [25]. An elegant prospective study demonstrated the untoward effects of increased total body water on patients who underwent colorectal surgery and who were randomized to a restrictive and a normal perioperative fluid regimen [26]. Patients who were randomized to the restrictive fluid regimen had significantly less complications, in particular cardiopulmonary complications, and better tissue healing.

Despite the fact that many AKI patients are already fluid overloaded, the majority of these patients receive fluid boluses in order to restore effective arterial blood volume and restore prerenal AKI. However, fluid overload may not only contribute to extra morbidity, but may also contribute to deterioration of kidney function. Especially in cardiac patients, increased central venous pressure and right ventricular failure is associated with the development of AKI [27–29]. Additionally, in acute respiratory distress syndrome, randomization to a restrictive fluid therapy regimen resulted in less need for RRT (10 vs. 14%, $p = 0.06$) [30]. Another mechanism by which fluid overload may lead to AKI is through the development of intra-abdominal hypertension, abdominal compartment syndrome, decreased thoracic and abdominal wall compliance, retroperitoneal edema, and ascites [25, 31, 32].

Inflammation and 'Organ Crosstalk'

AKI is characterized by a profound inflammatory reaction in the kidneys and in the systemic circulation. This systemic inflammatory response leads to dysfunction of other organs. Animal experiments have demonstrated that AKI leads to gene activation of proinflammatory and anti-inflammatory mediators in the lung, which results in acute respiratory distress syndrome with exudation of albumin in the alveoli, changes of aquaporins and sodium channels, and infiltration of neutrophils [33–40]. Acid-base disorders, which are commonly seen in AKI, also play a role in this inflammatory reaction. Hyperchloremic acidosis is associated with an increased interleukin (IL)-6/IL-10 ratio with a resulting proinflammatory effect, while lactic acidosis decreases both IL-6 and IL-10 resulting in an anti-inflammatory status [41–43]. More extensive discussions on this topic can be found in other chapters of this issue of *Contributions to Nephrology*.

Acidosis

The kidneys play an important role in acid-base homeostasis. Metabolic acidosis is the result of accumulation of anions such as chloride, phosphate and other anions that are not routinely measured [44, 45]. Acidosis occurs in up to one third of the patients who are initiated on RRT [46], interferes with normal functioning of many processes in the body, and leads to hemodynamic instability by decreased cardiac output and vasodilatation. Decreased density of β-receptors at the cell surface of the myocardium, interference with intracellular calcium handling, increased nitric oxide production and interference with the inflammatory response are the mechanisms that most likely play a role in this [41, 42, 47–49]. Further, different etiology of acidosis is associated with different systemic effects [41–43]. Moderate hyperchloremic acidosis is associated with increased nitric oxide production, leading to vasodilatation, while in lactic acidosis there is a gradual decline in nitric oxide production.

Electrolyte Abnormalities

The kidney also regulates electrolyte homeostasis. Up to one third of the patients with severe AKI will develop dilution hyponatremia by decreased free water clearance [50, 51]. Hyponatremia is associated with severe complications, such as cerebral edema, and worse outcomes [52, 53]. Furthermore, between 6.1% and one third of the patients who are initiated on RRT develop hyperkalemia, a condition that is associated with arrhythmias and death [46, 50].

Infection

Infection and antimicrobial therapy for infection play a central role in the course of AKI [54]. ICU patients with AKI have an increased prevalence of infection [54–58]. Infection and antimicrobial therapy may be the cause of AKI, but infection may also be a result of AKI. In a series at our center, we found that 80.2% of ICU patients who had AKI and who were treated with RRT were also treated for infection [55]. 37.5% of the patients even had two or more episodes of infection. Almost half of these infections started just before initiation of RRT, 40% during RRT and approximately 10% in the period immediately after discontinuation of RRT. Several factors may play a role in this interplay between infection, antimicrobial therapy and AKI.

Inadequate Antimicrobial Therapy

Adequate prescription of antimicrobial therapy is a challenge in ICU patients. The volume of distribution, metabolization and clearance can have important variations among patients and also within the same patient in different time periods of ICU stay. This may result in underdosing and overdosing of antimicrobials when standard antimicrobial dosing schedules are used.

Correct dosing is even more difficult in AKI patients. A first issue is correct evaluation of kidney function. Formulas that are used for assessment of kidney function in patients with chronic kidney disease, such as the Cockcroft-Gault and MDRD equations, were validated in non-ICU patients with moderate chronic kidney disease and are based on serum creatinine and variables such as age, body weight and gender. These equations are not adequate for assessment of kidney function in ICU patients [59, 60].

Kidney function is best assessed by measurement of urinary creatinine clearance, i.e. (urine volume × urine creatinine concentration)/(time in minutes × serum creatinine concentration). This calculation requires exact timing and measurement of urine volume, and a stable kidney function during the measurement period. As this condition is seldom met in patients with AKI, one can shorten the measurement period to, for example, 2 or 4 h or use the mean of serum creatinine concentration measured just before and after the measurement period [60].

Another issue that precludes correct dosing of antimicrobial therapy in ICU patients who have AKI is that dosing schedules for antimicrobial therapy are mostly based on data from patients with chronic kidney disease. These are not necessarily useful in ICU patients with AKI and a comparable degree of GFR. Serum concentration can be lower in ICU patients with increased volume of distribution, decreased gastrointestinal absorption, increased GFR during treatment, or RRT. Examples of factors that may increase serum concentrations are decreased albumin concentration, decreased kidney function and periods without RRT.

In patients treated with RRT, dosing schedules are available. But variables such as dialysis blood flow, dialysate flow, ultrafiltration rate, administration of pre- or postdilution, and filter characteristics may vary from center to center and have an impact on dialysis dose and clearance.

Inadequate Metabolic and Nutritional Support

ICU patients with AKI are usually in a catabolic state, and treatment with RRT leads to additional losses of amino acids and proteins. Loss of phosphorus in continuous RRT can lead to prolonged time on mechanical ventilation [61]. Further, the concentration of trace elements can be lower as a result of acute phase reaction, losses of fluids and removal by RRT. Finally, water-soluble vitamins such as vitamin C, thiamine and folic acid are removed by RRT [62].

At present, the data on the effects of nutritional interventions and different RRT modalities on nutritional status, and blood concentrations of trace elements and vitamins in ICU patients with AKI are insufficient. Given the data that we do know, and given the vast evidence on the importance of nutritional status and nutritional interventions in chronic hemodialysis patients and in ICU patients in general, this aspect of care needs further exploration.

Conclusion

Current epidemiologic findings demonstrate the strong association between AKI and short-term and long-term mortality. A whole range of clinical complications of AKI help to explain this. Factors that may help explain increased morbidity and mortality in AKI patients are a consequence of decreased kidney function such as volume overload, acidosis and electrolyte abnormalities. AKI may also impact on other organs, as in organ crosstalk between kidneys and lungs. AKI patients have an increased incidence of infection. Infection may impact on mortality, but also, inadequate antimicrobial therapy may play an important role. Current dosing recommendations for antimicrobials are mostly inadequate for ICU patients who have AKI, and adequate dosing is therefore a topic that needs further study. Finally, nutritional support is an underemphasized aspect of care for AKI patients in the ICU. Especially in AKI patients treated with RRT, we need more data on nutrition and supplementation of trace elements and vitamins.

The paradigm shift that patients die of, rather than with AKI, emphasizes the need for early recognition of AKI or clinical circumstances that eventually can lead to the development of AKI. RIFLE and AKIN criteria can be useful tools for intensivists in the early identification and management of AKI. The prevention of AKI in critically ill patients cannot be overemphasized. Adequate fluid therapy, the correction of acid-base disorders and electrolyte imbalances, the early recognition of infections, and adequate dosing of antimicrobial therapy are key issues in the management of AKI and in reducing its additional mortality in critically ill patients.

References

1 Kellum JA, Angus DC: Patients are dying of acute renal failure. Crit Care Med 2002;30:2156–2157.

2 Chertow GM, Levy EM, Hammermeister KE, Grover F, Daley J: Independent association between acute renal failure and mortality following cardiac surgery. Am J Med 1998;104:343–348.

3 Metnitz PG, Krenn CG, Steltzer H, et al: Effect of acute renal failure requiring renal replacement therapy on outcome in critically ill patients. Crit Care Med 2002;30:2051–2058.

4 Hoste EA, Lameire NH, Vanholder RC, Benoit DD, Decruyenaere JM, Colardyn FA: Acute renal failure in patients with sepsis in a surgical ICU: predictive factors, incidence, comorbidity, and outcome. J Am Soc Nephrol 2003;14:1022–1030.

5 Hoste EAJ, Schurgers M: Epidemiology of AKI: how big is the problem? Crit Care Med 2008;36:S1–S4.
6 Elseviers MM, Lins RL, Van der Niepen P, et al: Renal replacement therapy is an independent risk factor for mortality in critically ill patients with acute kidney injury. Crit Care 2010;14:R221.
7 Levy EM, Viscoli CM, Horwitz RI: The effect of acute renal failure on mortality. A cohort analysis. JAMA 1996;275:1489–1494.
8 Lassnigg A, Schmidlin D, Mouhieddine M, et al: Minimal changes of serum creatinine predict prognosis in patients after cardiothoracic surgery: a prospective cohort study. J Am Soc Nephrol 2004;15:1597–1605.
9 Lassnigg A, Schmid ER, Hiesmayr M, et al: Impact of minimal increases in serum creatinine on outcome in patients after cardiothoracic surgery: do we have to revise current definitions of acute renal failure? Crit Care Med 2008;36:1129–1137.
10 Chertow GM, Burdick E, Honour M, Bonventre JV, Bates DW: Acute kidney injury, mortality, length of stay, and costs in hospitalized patients. J Am Soc Nephrol 2005;16:3365–3370.
11 Liangos O, Wald R, O'Bell JW, Price L, Pereira A, Jaber BL: Epidemiology and outcomes of acute renal failure in hospitalized patients: a national survey. Clin J Am Soc Nephrol 2006;1:43–51.
12 Thakar CV, Christianson A, Freyberg R, Almenoff P, Render ML: Incidence and outcomes of acute kidney injury in intensive care units: a Veterans Administration study. Critical Care Medicine 2009;37:2552–2558.
13 Hoste EA, Clermont G, Kersten A, et al: RIFLE criteria for acute kidney injury are associated with hospital mortality in critically ill patients: a cohort analysis. Crit Care 2006;10:R73.
14 Ricci Z, Cruz D, Ronco C: The RIFLE criteria and mortality in acute kidney injury: a systematic review. Kidney Int 2008;73:538–546.
15 Brusselaers N, Monstrey S, Colpaert K, Decruyenaere J, Blot SI, Hoste EA: Outcome of acute kidney injury in severe burns: a systematic review and meta-analysis. Intensive Care Med 2010;36:915–925.
16 Ishani A, Xue JL, Himmelfarb J, et al: Acute kidney injury increases risk of ESRD among elderly. J Am Soc Nephrol 2009;20:223–228.
17 Coca SG, Peixoto AJ, Garg AX, Krumholz HM, Parikh CR: The prognostic importance of a small acute decrement in kidney function in hospitalized patients: a systematic review and meta-analysis. Am J Kidney Dis 2007;50:712–720.
18 Bihorac A, Yavas S, Subbiah S, et al: Long-term risk of mortality and acute kidney injury during hospitalization after major surgery. Ann Surg 2009;249:851–858.
19 Hobson CE, Yavas S, Segal MS, et al: Acute kidney injury is associated with increased long-term mortality after cardiothoracic surgery. Circulation 2009;119:2444–2453.
20 Lafrance JP, Miller DR: Acute kidney injury associates with increased long-term mortality. J Am Soc Nephrol 2010;21:345–352.
21 Logeart D, Tabet JY, Hittinger L, et al: Transient worsening of renal function during hospitalization for acute heart failure alters outcome. Int J Cardiol 2008;127:228–232.
22 Payen D, de Pont A, Sakr Y, et al: A positive fluid balance is associated with a worse outcome in patients with acute renal failure. Critical Care 2008;12:R74.
23 Bouchard J, Soroko SB, Chertow GM, et al: Fluid accumulation, survival and recovery of kidney function in critically ill patients with acute kidney injury. Kidney Int 2009;76:422–427.
24 Grams ME, Estrella MM, Coresh J, Brower RG, Liu KD: Fluid balance, diuretic use, and mortality in acute kidney injury. Clin J Am Soc Nephrol 2011;6:966–973.
25 De Waele JJ, De Laet I, Kirkpatrick AW, Hoste E: Intra-abdominal hypertension and abdominal compartment syndrome. Am J Kidney Dis 2011;57:159–169.
26 Brandstrup B, Tonnesen H, Beier-Holgersen R, et al: Effects of intravenous fluid restriction on postoperative complications: comparison of two perioperative fluid regimens: a randomized assessor-blinded multicenter trial. Ann Surg 2003;238:641–648.
27 Testani JM, Khera AV, St John Sutton MG, et al: Effect of right ventricular function and venous congestion on cardiorenal interactions during the treatment of decompensated heart failure. Am J Cardiol 2010;105:511–516.

28 Testani JM, Chen J, McCauley BD, Kimmel SE, Shannon RP: Potential effects of aggressive decongestion during the treatment of decompensated heart failure on renal function and survival. Circulation 2010;122:265–272.

29 Mullens W, Abrahams Z, Francis GS, et al: Importance of venous congestion for worsening of renal function in advanced decompensated heart failure. J Am Coll Cardiol 2009;53:589–596.

30 Wiedemann HP, Wheeler AP, Bernard GR, et al: Comparison of two fluid-management strategies in acute lung injury. N Engl J Med 2006;354:2564–2575.

31 Dalfino L, Tullo L, Donadio I, Malcangi V, Brienza N: Intra-abdominal hypertension and acute renal failure in critically ill patients. Intensive Care Med 2008;34:707–713.

32 De Waele JJ, De Laet I: Intra-abdominal hypertension and the effect on renal function. Acta Clin Belg Suppl 2007;371–374.

33 Grigoryev DN, Liu M, Hassoun HT, Cheadle C, Barnes KC, Rabb H: The local and systemic inflammatory transcriptome after acute kidney injury. J Am Soc Nephrol 2008;19:547–558.

34 Hassoun HT, Grigoryev DN, Lie ML, et al: Ischemic acute kidney injury induces a distant organ functional and genomic response distinguishable from bilateral nephrectomy. Am J Physiol Renal Physiol 2007;293:F30–F40.

35 Hoke TS, Douglas IS, Klein CL, et al: Acute renal failure after bilateral nephrectomy is associated with cytokine-mediated pulmonary injury. J Am Soc Nephrol 2007;18:155–164.

36 Kim DJ, Park SH, Sheen MR, et al: Comparison of experimental lung injury from acute renal failure with injury due to sepsis. Respiration 2006;73:815–824.

37 Ko GJ, Rabb H, Hassoun HT: Kidney-lung crosstalk in the critically ill patient. Blood Purif 2009;28:75–83.

38 Kramer AA, Postler G, Salhab KF, Mendez C, Carey LC, Rabb H: Renal ischemia/reperfusion leads to macrophage-mediated increase in pulmonary vascular permeability. Kidney Int 1999;55:2362–2367.

39 Rabb H, Chamoun F, Hotchkiss J: Molecular mechanisms underlying combined kidney-lung dysfunction during acute renal failure; in Ronco C, Bellomo R, La Greca G (eds): Blood Purification in Intensive Care. Contrib Nephrol. Basel, Karger, 2001, pp 41–52.

40 Rabb H, Wang Z, Nemoto T, Hotchkiss J, Yokota N, Soleimani M: Acute renal failure leads to dysregulation of lung salt and water channels. Kidney Int 2003;63:600–606.

41 Kellum JA, Song M, Venkataraman R: Effects of hyperchloremic acidosis on arterial pressure and circulating inflammatory molecules in experimental sepsis. Chest 2004;125:243–248.

42 Kellum JA, Song M, Li J: Science review: extracellular acidosis and the immune response: clinical and physiologic implications. Crit Care 2004;8:331–336.

43 Kellum JA, Song M, Li J: Lactic and hydrochloric acids induce different patterns of inflammatory response in LPS-stimulated RAW 264.7 cells. Am J Physiol Regul Integr Comp Physiol 2004;286:R686–R692.

44 Rocktaeschel J, Morimatsu H, Uchino S, et al: Acid-base status of critically ill patients with acute renal failure: analysis based on Stewart-Figge methodology. Crit Care 2003;7:R60.

45 Forni LG, McKinnon W, Lord GA, Treacher DF, Peron JM, Hilton PJ: Circulating anions usually associated with the Krebs cycle in patients with metabolic acidosis. Crit Care 2005;9:R591–R595.

46 Bellomo R, Cass A, Cole L, et al: Intensity of continuous renal-replacement therapy in critically ill patients. N Engl J Med 2009;361:1627–1638.

47 Kellum JA, Song M, Almasri E: Hyperchloremic acidosis increases circulating inflammatory molecules in experimental sepsis. Chest 2006;130:962–967.

48 Marsh JD, Margolis TI, Kim D: Mechanism of diminished contractile response to catecholamines during acidosis. Am J Physiol 1988;254:H20–H27.

49 Nimmo AJ, Than N, Orchard CH, Whitaker EM: The effect of acidosis on beta-adrenergic receptors in ferret cardiac muscle. Exp Physiol 1993;78:95–103.

8 Hassoun HT, Lie ML, Grigoryev DN, Liu M, Tuder RM, Rabb H: Kidney ischemia-reperfusion injury induces caspase-dependent pulmonary apoptosis. Am J Physiol Renal Physiol 2009;297:F125–F137.
9 Zarbock A, Schmolke M, Spieker T, Jurk K, Van Aken H, Singbartl K: Acute uremia but not renal inflammation attenuates aseptic acute lung injury: a critical role for uremic neutrophils. J Am Soc Nephrol 2006;17:3124–3131.
10 Kim do J, Park SH, Sheen MR, Jeon US, Kim SW, Koh ES, Woo SK: Comparison of experimental lung injury from acute renal failure with injury due to sepsis. Respiration 2006;73:815–824.
11 Rossaint J, Spelten O, Kaessens K, Mueller H, Van Aken H, Singbartl K, Zarbock A: Acute loss of renal function abolishes slow leukocyte rolling and transmigration by interfering with intracellular-signaling molecules. Kidney Int 2011, DOI: 10.1038/ki.2011.125.
12 Tsai WC, Strieter RM, Mehrad B, Newstead MW, Zeng X, Standiford TJ: CXC chemokine receptor CXCR2 is essential for protective innate host response in murine *Pseudomonas aeruginosa* pneumonia. Infect Immun 2000;68:4289–4296.
13 De Waele JJ, Hoste EA, Blot SI: Blood stream infections of abdominal origin in the intensive care unit: characteristics and determinants of death. Surg Infect (Larchmt) 2008;9:171–177.
14 Hoste EA, Blot SI, Lameire NH, Vanholder RC, De Bacquer D, Colardyn FA: Effect of nosocomial bloodstream infection on the outcome of critically ill patients with acute renal failure treated with renal replacement therapy. J Am Soc Nephrol 2004;15:454–462.
15 Grigoryev DN, Liu M, Hassoun HT, Cheadle C, Barnes KC, Rabb H: The local and systemic inflammatory transcriptome after acute kidney injury. J Am Soc Nephrol 2008;19:547–558.
16 Tracz MJ, Juncos JP, Croatt AJ, et al: Deficiency of heme oxygenase-1 impairs renal hemodynamics and exaggerates systemic inflammatory responses to renal ischemia. Kidney Int 2007;72:1073–1080.
17 Hassoun HT, Grigoryev DN, Lie ML, Liu M, Cheadle C, Tuder RM, Rabb H: Ischemic acute kidney injury induces a distant organ functional and genomic response distinguishable from bilateral nephrectomy. Am J Physiol Renal Physiol 2007;293:F30–F40.
18 Kuiper JW, Groeneveld ABJ, Slutsky AS, Plötz FB: Mechanical ventilation and acute renal failure. Crit Care Med 2005;33:1408–1415.
19 Murdaugh HV, Sieker HO, Manfredi F: Effect of altered intrathoracic pressure on renal hemodynamics, electrolyte excretion and water clearance. J Clin Invest 1959;38:834–842.
20 Hall SV, Johnson EE, Hedley-Whyte J: Renal hemodynamics and function with continuous positive-pressure ventilation in dogs. Anesthesiology 1974;41:452–461.
21 Ventilation with lower tidal volumes as compared with traditional tidal volumes for acute lung injury and the acute respiratory distress syndrome. The Acute Respiratory Distress Syndrome Network. N Engl J Med 2000;342:1301–1308.
22 Ranieri VM, Suter PM, Tortorella C, et al: Effect of mechanical ventilation on inflammatory mediators in patients with acute respiratory distress syndrome: a randomized controlled trial. JAMA 1999; 282:54–61.
23 Ranieri VM, Giunta F, Suter PM, Slutsky AS: Mechanical ventilation as a mediator of multisystem organ failure in acute respiratory distress syndrome. JAMA 2000;284:43–44.
24 Murugan R, Karajala-Subramanyam V, Lee M, et al: Acute kidney injury in nonsevere pneumonia is associated with an increased immune response and lower survival. Kidney Int 2010;77:527–535.

Kai Singbartl, MD, MPH
Department of Critical Care Medicine
University of Pittsburgh
3550 Terrace Street, Pittsburgh, PA 15261 (USA)
Tel. +1 412 647 1473, E-Mail kas148@pitt.edu

Kellum JA, Ronco C, Vincent J-L (eds): Controversies in Acute Kidney Injury.
Contrib Nephrol. Basel, Karger, 2011, vol 174, pp 71–77

Acute Kidney Injury, Acute Lung Injury and Septic Shock: How Does Mortality Compare?

Jean-Louis Vincent

Department of Intensive Care, Erasme Hospital, Université libre de Bruxelles, Bruxelles, Belgium

Abstract

Acute kidney injury (AKI), acute lung injury (ALI) and sepsis are all commonly encountered in critically ill patients. Although considered as separate conditions, largely for therapeutic purposes, a common inflammatory response is often implicated in their pathophysiologies and they are frequently present simultaneously. Mortality rates in critically ill patients suffering from renal failure, respiratory failure or severe sepsis are quite similar at about 40%, and all increase substantially when these conditions coexist. Most intensive care unit patients will die from multiple rather than individual organ failure, and further research is needed to evaluate the patterns of organ failure in surviving and nonsurviving critically ill patients, as well as the importance and mechanisms of organ-organ crosstalk in such patients.

Acute kidney injury (AKI), acute lung injury (ALI) and sepsis are distinct entities, but in critically ill patients they are often inextricably linked. Sepsis is the leading cause of both AKI and ALI in the intensive care unit (ICU) population, and a diagnosis of either AKI or ALI makes a patient more likely to develop sepsis. These processes must involve some similar mechanisms, and an inflammatory response is clearly implicated. We are beginning to unravel the important underlying processes whereby dysfunction in one organ influences another (so-called 'crosstalk') and starting to imagine how treatment of one organ dysfunction can improve function in a distant organ. Importantly, although definitions of ALI/ARDS (acute respiratory distress syndrome) have been fairly standardized since 1994 [1], definitions of acute renal failure/AKI have been fairly variable, although the new RIFLE/AKIN definitions are trying to resolve this, and similarly sepsis definitions also vary considerably despite the 2001 sepsis definitions conference [2]. For all these conditions, we lack clear markers to aid diagnosis and have to rely on a combination of clinical signs and available biochemical and radiological variables, rendering the development of clear definitions difficult. Study populations also vary in age, disease severity, reasons for admission, comorbidities, etc. – all factors that can influence outcomes. Data from different studies can therefore be difficult to compare, and this can be particularly tricky when trying to assess whether mortality rates have changed over time. Finally, many ICU patients will have a combination of ALI with AKI or sepsis or even all three conditions, and it is difficult to separate and compare individual mortality rates for the three entities.

Mortality from Acute Kidney Injury

AKI affects some 22–65% of ICU patients [3–7], the precise figure varying according to the definition used. The prognosis of patients with acute kidney failure remains poor despite considerable progress in renal replacement therapies (RRT). In a review on outcomes from renal failure, mortality remained around 50% during the period from 1970 to 2004 [8] (fig. 1), and this is unlikely to have changed substantially over the last few years. In a recent UK study of 821 patients who received RRT for AKI, the ICU mortality rate was 55% and the hospital mortality rate was 66% [9]. Similarly, a Belgian study reported hospital mortality rates of 43% in patients with AKI who did not require RRT and 58% in those who did [10].

Importantly, critically ill patients with AKI have higher mortality rates than ICU patients who do not develop renal failure [11, 12]. One may wonder why this is the case when RRT can be used to support the failing organ? Clearly, organ failure is a marker of the severity of the disease as shown by the increasing mortality rates with higher renal sequential organ failure assessment (SOFA) scores [13]. RRT is unable to entirely compensate for the failing kidneys. Moreover, mortality is largely related to the presence of other conditions and organ failures. In one early report from the EDTA-European Renal Association, patients treated on renal units with acute renal failure alone had mortality rates of just 8%, but this figure rose to 65% in combination with sepsis or liver failure, 71% with respiratory failure and 76% with circulatory failure [14]. Similar data have also been reported for critically ill patients. In the European SOAP study in ICU patients [15], patients with renal failure at any time had hospital mortality rates of 12%, while the mortality rates in patients who also had other organ failures was much higher (20% in patients with concomitant respiratory failure, 43% in patients with cardiovascular failure, and 53% in patients with central nervous system failure; fig. 2). In a retrospective analysis of data from the Australian

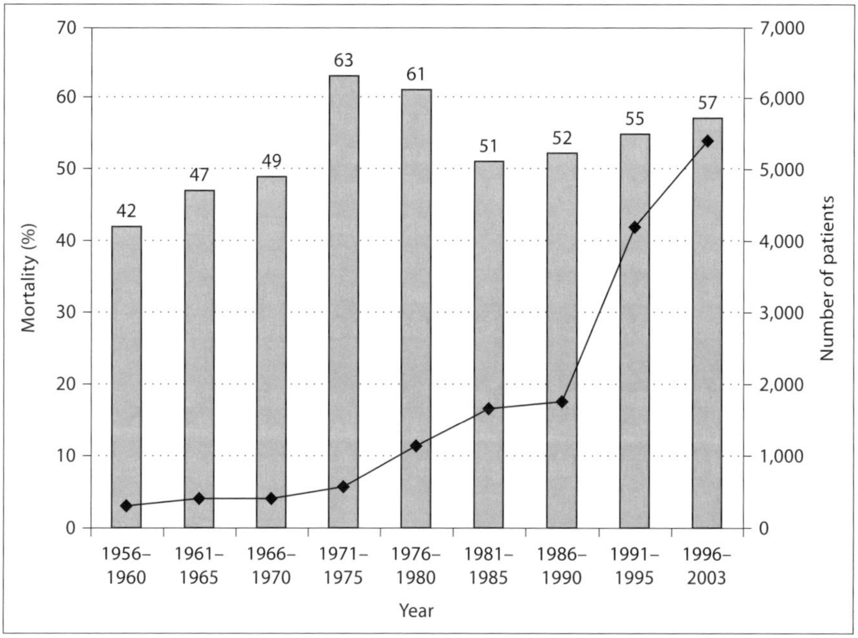

Fig. 1. Changes in mortality (bar graph) in patients with acute renal failure over the 47 years between 1956 and 2003. Line represents numbers of patients included in the studies during each time period. From Ympa et al. [8] with permission.

and New Zealand Intensive Care Society Adult Patient Database, patients with septic AKI had significantly larger hospital mortality rates than patients with nonseptic AKI (29.7 vs. 21.6%, $p < 0.001$); interestingly, these differences were similar regardless of the severity of the AKI [16].

Mortality from Severe Sepsis

Up to 30% of ICU patients will have severe sepsis at some point during their ICU stay [15], and the incidence is increasing [17, 18]. Mortality rates for patients with severe sepsis are in the range of 30–40% [15]. The PROWESS (Protein C Worldwide Evaluation in Severe Sepsis) study, in which the mortality rate was around 30% in the placebo group, is often taken as a reference for mortality rates in severe sepsis [19], but clinical trials for new therapeutic interventions have many exclusion criteria, notably for patients with a poor prognosis (e.g. patients with advanced cancer, cirrhosis and immunosuppression, or presence of a 'do-not-resuscitate' order are excluded). Moreover as PROWESS was a trial of a drug with anticoagulant properties, patients with a high risk of bleeding were also excluded, and these patients obviously have a poorer prognosis. Hence, the

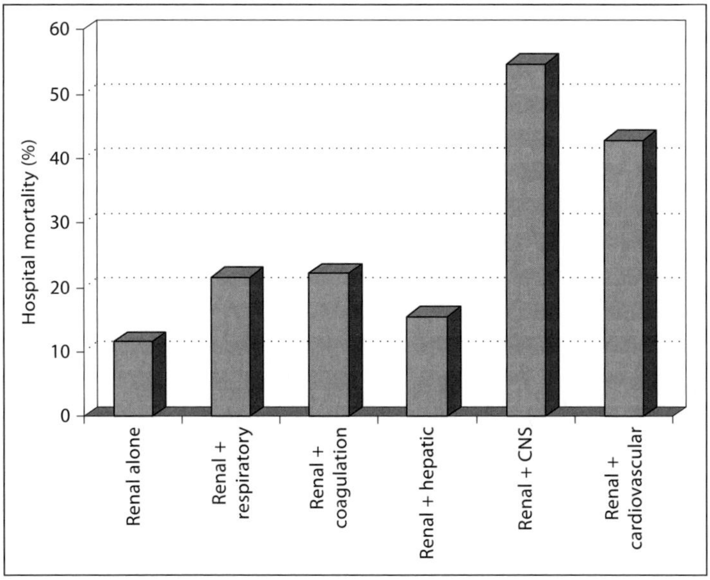

Fig. 2. Hospital mortality rates in patients in the SOAP study with isolated renal failure on admission or a combination of renal and other organ failures.

mortality from severe sepsis in the real-life clinical situation is expected to be higher than the 30% reported in PROWESS. Observational studies can provide more reliable real-life data on general mortality rates than randomized controlled studies, and results from several recent epidemiological studies report hospital mortality rates ranging from 28% to as high as 55% for patients with severe sepsis [15, 18, 20–25]. Rates are even higher for patients with septic shock. Differences among studies are largely the result of different study populations, e.g. older patients will have higher mortality rates than younger patients.

Mortality from Acute Respiratory Failure

ALI/ARDS affects 15–25% of ICU patients [26] depending on the population studied, with higher rates if only mechanically ventilated patients are included [27]. Mortality rates are around 30–45% [26–29], again depending on the definition used and the study population. There may have been some improvement in survival rates over time, with Zambon and Vincent [28] reporting a decrease in hospital mortality rates between 1994 and 2006. Another meta-analysis found no similar decrease [29], but it had methodological limitations [30]. As in AKI, sepsis frequently coexists in patients with ALI/ARDS. Several studies have reported higher mortality rates in patients with septic ALI/ARDS compared to nonseptic ALI/ARDS, but this increase seems to be largely due to the greater disease severity and other organ dysfunctions present in such patients rather than a direct effect of the presence of sepsis alone [31, 32]. Again similar to the situation with AKI, mortality rates are strongly influenced by the simultaneous presence of other organ failures, with survival decreasing with the number of failing organs [33, 34].

Mortality from Multiple Organ Failure

As seen above, AKI, ALI and severe sepsis all have high mortality rates, but these rates are largely linked to the association of the individual syndromes with dysfunction of other organs, leading ultimately to multiple organ failure. In all the above conditions – AKI, ALI and sepsis – mortality rates are higher when there is additional organ failure. Indeed, most patients who die of AKI, ALI or severe sepsis do so because of multiple organ failure rather than as a direct result of their kidney dysfunction or hypoxemia [35, 36]. In a recent analysis of the INDEPTH database, which combines data from severe sepsis patients who were included in 5 clinical trials performed between July 1996 and December 2002, we reported that 43% of patients died from sepsis-associated multiple organ failure. Importantly, although 80% of nonsurvivors had an increase in at least one individual organ SOFA score in the 4 days prior to death, the mean changes in individual scores were small and the most common pattern over the days preceding death was one of persistent multiple organ function rather than any dramatic worsening of organ dysfunction [36]. The development of multiple organ failure may lead to withdrawing or withholding of therapy, now a common event preceding death in ICU patients. In the SOAP study, we showed that more patients developed renal dysfunction in the day prior to death than developed other organ dysfunctions, suggesting that the onset of renal dysfunction may be a strong determinant of the decision to withhold or withdraw life-supporting therapy. The patterns of organ dysfunction during ICU admission in survivors and nonsurvivors need to be further evaluated.

Conclusion

Renal failure is common in acutely ill patients and associated with high mortality rates, despite the considerable advances in RRT in recent years. Interestingly, mortality rates from renal failure, respiratory failure and severe sepsis are quite similar, and all increase substantially when these conditions coexist.

References

1 Bernard GR, Artigas A, Brigham KL, et al: The American-European Consensus Conference on ARDS. Definitions, mechanisms, relevant outcomes, and clinical trial coordination. Am J Respir Crit Care Med 1994;149:818–824.

2 Levy MM, Fink MP, Marshall JC, et al: 2001 SCCM/ESICM/ACCP/ATS/SIS International Sepsis Definitions Conference. Crit Care Med 2003;31:1250–1256.

3 de Mendonca A, Vincent JL, Suter PM, et al: Acute renal failure in the ICU: risk factors and outcome evaluated by the SOFA score. Intensive Care Med 2000;26:915–921.

4 Hoste EA, Clermont G, Kersten A, et al: RIFLE criteria for acute kidney injury are associated with hospital mortality in critically ill patients: a cohort analysis. Crit Care 2006;10:R73.

5 Ostermann M, Chang RW: Acute kidney injury in the intensive care unit according to RIFLE. Crit Care Med 2007;35:1837–1843.

6 Thakar CV, Christianson A, Freyberg R, et al: Incidence and outcomes of acute kidney injury in intensive care units: a Veterans Administration study. Crit Care Med 2009;37:2552–2558.

7 Bagshaw SM, George C, Dinu I, et al: A multi-centre evaluation of the RIFLE criteria for early acute kidney injury in critically ill patients. Nephrol Dial Transplant 2008;23:1203–1210.

8 Ympa YP, Sakr Y, Reinhart K, et al: Has mortality from acute renal failure decreased? A systematic review of the literature. Am J Med 2005;118:827–832.

9 Lines SW, Cherukuri A, Murdoch SD, et al: The outcomes of critically ill patients with acute kidney injury receiving renal replacement therapy. Int J Artif Organs 2011;34:2–9.

10 Elseviers MM, Lins RL, Van der NP, et al: Renal replacement therapy is an independent risk factor for mortality in critically ill patients with acute kidney injury. Crit Care 2010;14:R221.

11 Garzotto F, Piccinni P, Cruz D, et al: RIFLE-based data collection/management system applied to a prospective cohort multicenter Italian study on the epidemiology of acute kidney injury in the intensive care unit. Blood Purif 2011;31:159–171.

12 Hoste EA, Schurgers M: Epidemiology of acute kidney injury: how big is the problem? Crit Care Med 2008;36:S146–S151.

13 Vincent JL, Moreno R, Takala J, et al: The SOFA (sepsis-related organ failure assessment) score to describe organ dysfunction/failure. On behalf of the Working Group on Sepsis-Related Problems of the European Society of Intensive Care Medicine. Intensive Care Med 1996;22:707–710.

14 Brunner F, Broyer M, Brynger H, et al: Combined report on regular dialysis and transplantation in Europe XV, 1984. Proc Eur Dial Transplant Assoc 1985;22:3–53.

15 Vincent JL, Sakr Y, Sprung CL, et al: Sepsis in European intensive care units: results of the SOAP study. Crit Care Med 2006;34:344–353.

16 Bagshaw SM, George C, Bellomo R: Early acute kidney injury and sepsis: a multicentre evaluation. Crit Care 2008;12:R47.

17 Brun-Buisson C, Meshaka P, Pinton P, et al: EPISEPSIS: a reappraisal of the epidemiology and outcome of severe sepsis in French intensive care units. Intensive Care Med 2004;30:580–588.

18 Dombrovskiy VY, Martin AA, Sunderram J, et al: Rapid increase in hospitalization and mortality rates for severe sepsis in the United States: a trend analysis from 1993 to 2003. Crit Care Med 2007;35:1244–1250.

19 Bernard GR, Vincent JL, Laterre PF, et al: Efficacy and safety of recombinant human activated protein C for severe sepsis. N Engl J Med 2001;344:699–709.

20 Engel C, Brunkhorst FM, Bone HG, et al: Epidemiology of sepsis in Germany: results from a national prospective multicenter study. Intensive Care Med 2007;33:606–618.

21 Blanco J, Muriel-Bombin A, Sagredo V, et al: Incidence, organ dysfunction and mortality in severe sepsis: a Spanish multicentre study. Crit Care 2008;12:R158.

22 Malacarne P, Langer M, Nascimben E, et al: Building a continuous multicenter infection surveillance system in the intensive care unit: findings from the initial data set of 9,493 patients from 71 Italian intensive care units. Crit Care Med 2008;36:1105–1113.

23 Shen HN, Lu CL, Yang HH: Epidemiologic trend of severe sepsis in Taiwan from 1997 through 2006. Chest 2010;138:298–304.
24 Martin G, Brunkhorst FM, Janes JM, et al: The international PROGRESS registry of patients with severe sepsis: drotrecogin alfa (activated) use and patient outcomes. Crit Care 2009;13:R103.
25 Cheng B, Xie G, Yao S, et al: Epidemiology of severe sepsis in critically ill surgical patients in ten university hospitals in China. Crit Care Med 2007;35:2538–2546.
26 Irish Critical Care Trials Group: Acute lung injury and the acute respiratory distress syndrome in Ireland: a prospective audit of epidemiology and management. Crit Care 2008;12:R30.
27 Rubenfeld GD, Caldwell E, Peabody E, et al: Incidence and outcomes of acute lung injury. N Engl J Med 2005;353:1685–1693.
28 Zambon M, Vincent JL: Mortality rates for patients with acute lung injury/ARDS have decreased over time. Chest 2008;133:1120–1127.
29 Phua J, Badia JR, Adhikari NK, et al: Has mortality from acute respiratory distress syndrome decreased over time? A systematic review. Am J Respir Crit Care Med 2009;179:220–227.
30 Zambon M, Vincent JL: Are outcomes improving in patients with ARDS? Am J Respir Crit Care Med 2009;180:1158–1159.
31 Sevransky JE, Martin GS, Shanholtz C, et al: Mortality in sepsis versus non-sepsis induced acute lung injury. Crit Care 2009;13:R150.
32 Sheu CC, Gong MN, Zhai R, et al: Clinical characteristics and outcomes of sepsis-related vs non-sepsis-related ARDS. Chest 2010;138:559–567.
33 Flaatten H, Gjerde S, Guttormsen AB, et al: Outcome after acute respiratory failure is more dependent on dysfunction in other vital organs than on the severity of the respiratory failure. Crit Care 2003;7:R72.
34 Aggarwal AN, Agarwal R, Gupta D, et al: Nonpulmonary organ dysfunction and its impact on outcome in patients with acute respiratory failure. Chest 2007;132:829–835.
35 Mayr VD, Dunser MW, Greil V, et al: Causes of death and determinants of outcome in critically ill patients. Crit Care 2006;10:R154.
36 Vincent JL, Nelson DR, Williams MD: Is worsening multiple organ failure the cause of death in patients with severe sepsis? Crit Care Med 2011;39:1050–1055.

Dr. Jean-Louis Vincent
Department of Intensive Care
Erasme University Hospital
Route de Lennik 808, BE–1070 Brussels (Belgium)
Tel. +32 2 555 3380, E-Mail jlvincen@ulb.ac.be

Kellum JA, Ronco C, Vincent J-L (eds): Controversies in Acute Kidney Injury.
Contrib Nephrol. Basel, Karger, 2011, vol 174, pp 78–88

Sepsis and Acute Kidney Injury Are Bidirectional

Martin Matejovic · Jiri Chvojka · Jaroslav Radej · Lenka Ledvinova · Thomas Karvunidis · Ales Krouzecky · Ivan Novak

ICU, 1st Medical Department, Charles University in Prague, Faculty of Medicine in Plzen, University Hospital Plzen, Plzen, Czech Republic

Abstract

Sepsis is the most common cause of acute kidney injury (AKI). There has been a growing body of evidence demonstrating the association between worsening of kidney function during sepsis and the risk of short- and long-term mortality. AKI in sepsis is associated with poor outcome and independently predicts increased mortality. Sepsis-associated AKI may therefore serve as a biomarker of adverse physiological events that portends worse outcome. Conversely, the important role of sepsis among intensive care unit patients with nonseptic AKI is increasingly being recognized. Indeed, sepsis represents a significant contributing factor to the overall mortality and incomplete recovery of kidney function in subjects who developed nonseptic AKI. Because AKI portends such an ominous prognosis in sepsis and vice versa, there has been a surge of interest in elucidating mechanisms underlying the complex and bidirectional nature of the interconnections between AKI, sepsis and multiorgan dysfunction. Accumulating data indicate that AKI can trigger several immune, metabolic and humoral pathways, thus potentially contributing to distant organ dysfunction and overall morbidity and mortality. The expanding population of patients with sepsis and AKI, and the associated excess mortality provide a strong basis for further research aimed at addressing more rigorously all potentially modifiable factors to reduce this burden to patients and health care systems. Better insights into bidirectional and synergistic pathways linking sepsis and AKI might open the window for new therapeutic approaches that interrupt this vicious circle. Here, we discuss the rationale for and the current understanding of the bidirectional relationship between AKI and sepsis.

The burden of sepsis and acute kidney injury (AKI) is steadily increasing in critically ill patients, and both portend an ominous outcome [1]. Sepsis is the dominant cause of AKI, accounting for nearly 50% of episodes of acute renal failure [2]. When occurring together, the associated mortality is significantly higher as compared to nonseptic AKI or sepsis alone [3]. Conversely, the important role of sepsis among intensive care unit (ICU) patients with nonseptic AKI is increasingly being recognized. Although many existing studies have important methodological issues that hamper their ability to unambiguously determine the independent contribution of AKI to these outcomes, both basic and clinical research data suggest there may be a bidirectional and synergistic relationship between sepsis and AKI that magnifies the poor outcomes associated with either disease alone.

Sepsis and Acute Kidney Injury: Clinical Problem

Epidemiology of Acute Kidney Injury in Sepsis

The major importance of sepsis and AKI as a clinical and public health issue has clearly been underscored by recent studies showing a steadily rising incidence of both syndromes [1, 4]. Sepsis and septic shock are the most common causes of mortality in noncoronary ICUs, accounting for 37% of ICU admissions [5]. Similarly, AKI is a frequently encountered problem in critically ill patients, occurring in 36% of all ICU patients [6]. Although AKI is often a complex multifactorial syndrome, sepsis and septic shock are its dominant causes in this patient population [2, 3]. According to several epidemiological studies, 11–60% of septic patients have concomitant AKI, depending on the definition used and population studied (reviewed in [2]). Moreover, the incidence of AKI proportionally rises with the severity of sepsis, occurring in 19% of patients with sepsis, 23% patients with severe sepsis and 51–64% patients with septic shock [3, 7, 8]. Therefore, sepsis-associated AKI (SA-AKI) defines the simultaneous presence of both syndromes – sepsis and AKI – in the absence of other clear and established precipitants of AKI, and is an exceedingly common and important issue in the ICU [2].

Epidemiology of Sepsis in Nonseptic Acute Kidney Injury

AKI not only complicates the course of sepsis, but it also appears to predispose patients to the development of sepsis [9]. For example, in patients with contrast-induced AKI who subsequently died, 45% developed sepsis as a complication of kidney injury [10]. In another study, a significantly higher incidence of infections (59 vs. 24%, $p < 0.001$) occurred in patients with AKI compared to patients without AKI after cardiovascular surgery [11]. Similarly, a greater proportion of AKI patients requiring renal replacement therapy developed blood stream infections when compared to patients without AKI (8.8 vs. 3.5%, $p < 0.001$) [12].

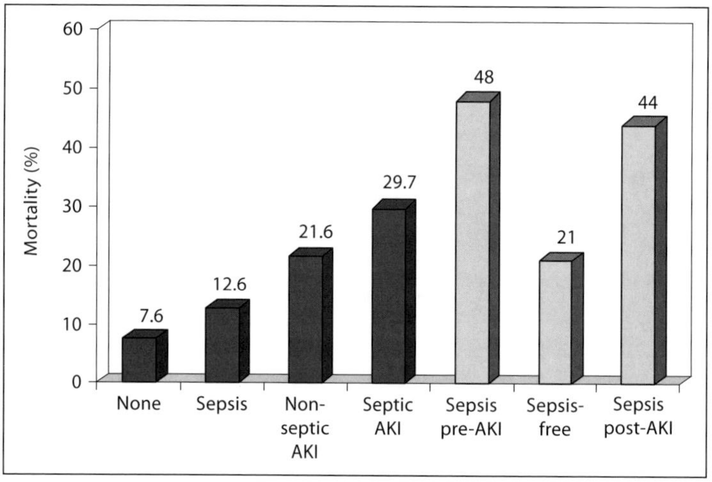

Fig. 1. Crude hospital mortality stratified by the subgroups (black columns, adapted from [18]); hospital mortality rate stratified for sepsis pre-AKI, sepsis-free AKI and sepsis post-AKI (grey columns, adapted from [13]).

Nevertheless, little is known about the role of sepsis among ICU patients with nonseptic AKI. This issue has recently been addressed by Mehta et al. [13], who studied the frequency with which sepsis develops as a complication of AKI and the clinical consequences of this sepsis. They analyzed data from 618 critically ill patients enrolled in a multicenter observational study of AKI (PICARD), and compared clinical characteristics of patients who remained sepsis-free with patients who developed sepsis post-AKI and with patients who developed sepsis prior to AKI diagnosis. Forty percent of the patients who were not septic at the time of AKI diagnosis developed sepsis during their hospital stay, and half of them within 5 days of their AKI diagnosis. These data suggest an increased susceptibility of patients with nonseptic AKI to infections and sepsis [14].

Impact of Septic Acute Kidney Injury on Clinical Outcomes

AKI has been shown to be a potent predictor of morbidity and mortality in a number of clinical settings, and extends beyond ICU and in-hospital adverse outcomes to long-term health consequences [15–17]. This phenomenon is even more pronounced when sepsis and AKI combine in critically ill patients (fig. 1). Moreover, a graded relationship between the severity of SA-AKI and mortality has been reported [8, 18]. Numerous observational studies have demonstrated that SA-AKI tends to systematically differ from those with nonseptic AKI or sepsis alone on relevant demographic and clinical characteristics as well as on several important outcomes (reviewed in [2]). The latter includes higher mortality, longer duration of hospital stay and trends towards greater recovery of renal functions to independence from renal replacement therapy [2]. Even among patients with nonsevere pneumonia, AKI was a common complication (16–25%) and continued to be associated with increased risk of death (hazard ratio of 1.29 in patients not requiring ICU care) [19].

Given these differences in clinically meaningful outcomes, accurate assessment of demographic, clinical and pathophysiological characteristics is paramount to delineating the risk attributable to SA-AKI for poor outcomes. In this context, compared to patients with nonseptic AKI, septic AKI patients have a higher burden of pre-existing comorbidities, higher acuity of illness and greater abnormalities in acute physiology, thus making the discrimination between the two entities clinically relevant [2]. This reasoning is further underscored by emerging experimental evidence suggesting that pathogenesis of septic AKI also involves distinct mechanisms as compared to nonseptic causes of AKI [20, 21].

Little is known about the long-term consequences of patients with SA-AKI once they leave the hospital. Nevertheless, a recent study by Lopes et al. [22] showed 3.2-fold risk of death of these patients during a 2-year observational period despite having greater likelihood for renal function recovery. Similarly, Murugan et al. [19] showed that the increased risk of death associated with AKI in patients with nonsevere community-acquired pneumonia persisted long after discharge (up to 100 days) from hospital. Although the mechanisms underlying negative long-term consequences after an episode of SA-AKI are not known, it is tempting to speculate that AKI as a complication of sepsis could identify a high-risk group of patients who failed a 'stress test', thus predisposing them to high rates of both short- and long-term mortality and morbidity [23]. Collectively, these data imply SA-AKI by itself adds to the morbidity and mortality of septic patients, thus representing an important driving force accelerating the downward spiral toward bad outcomes.

Impact of Sepsis on Clinical Outcomes of Nonseptic Acute Kidney Injury

As suggested, nonseptic AKI is often complicated by infection and sepsis [9]. In a cohort analytical study, sepsis was recognized as a significant contributing factor to the overall mortality in subjects who developed contrast-induced AKI and subsequently died (45%, 21/47) [10]. More detailed analysis of the risk factors and prognosis of sepsis developing after AKI has recently been published by Mehta et al. [13]. As mentioned above, 40% of the patients developed sepsis after AKI diagnosis. These patients, as opposed to AKI patients without sepsis, were more oliguric at AKI diagnosis, had a higher severity of illness and were less likely to have pre-existing chronic kidney disease. Independent predictors for the development of sepsis post-AKI, assessed by logistic regression, include oliguria ≥3 days, persistent fluid overload post-AKI diagnosis, invasive nonsurgical procedures, high severity of illness and dialysis requirement. The in-hospital mortality was significantly higher in patients with sepsis post-AKI than in sepsis-free patients (44 vs. 21%, $p < 0.001$) and similar to that of patients

Lessons Learned from End-Stage Renal Disease Patients. Acute infections contribute substantially to the high rates of hospitalization and mortality in end-stage renal disease patients. Infection is second leading cause of death in patients with end-stage renal disease, and sepsis accounts for more than 75% of these infectious deaths [29]. The reasons for this susceptibility include advanced age, severe comorbidities (e.g. diabetes mellitus, chronic heart disease), acquired immunosuppression related to uremia and increased exposure of patients to infectious/sepsis risk factors (e.g. dialysis access-related infections, procedure-related bioincompatibility) [29]. While these associations might not directly be transferable to patients with AKI, it is increasingly clear that the large population of AKI patients experiences similar alterations in homeostasis as well as a high rate of morbidity and mortality.

Factors Related to Abnormal Physiology Caused by Acute Kidney Injury. AKI leads to a pathologic milieu that adds the stress of fluid and electrolyte imbalance, waste clearance and immune dysfunction. A positive cumulative fluid balance has increasingly been recognized as an important and potentially modifiable determinant of clinical outcome in patients with AKI [30, 31]. In particular, patients with SA-AKI have been shown to have higher rates of oliguria despite having received more fluid therapy when compared with nonseptic AKI or those with sepsis alone [2]. Potential mechanisms by which excessive fluid therapy might negatively affect the course of critical illness might include increased risk for nosocomial infections and sepsis, impaired wound healing, interstitial edema-aggravated organ dysfunction, and delayed recovery of organ functions [13, 14, 32]. Furthermore, toxic effects of retained organic wastes, i.e. uremic toxins, involving oxidative stress, inflammation and insulin resistance may also contribute to the high rate of morbidity and mortality among septic patients [33]. On the other hand, recent experimental studies have indicated that it is probably kidney injury, not uremia per se, which constitutes the critical factor in mediating distant organ dysfunction [34].

Factors Related to Side Effects of Renal Replacement Therapy. The deleterious effects of AKI on the outcome of septic patients might also be attributable to the side effects of renal replacement therapy. In the EPISEPSIS study, for example, the need for renal replacement therapy was associated with an increased risk of death [35]. Although data indicating a causal relationship are lacking, dialysis-induced inflammatory response including leukocyte and complement activation, depletion of antioxidants, and dialysis access-related infections may assume biological plausibility in this context [36].

Molecular Mechanisms of AKI-induced Systemic Injury. Molecular mechanisms for the crosstalk between AKI and sepsis are poorly understood. Nevertheless, several lines of basic scientific evidence provide a pathophysiological rationale for a causal relationship between AKI and worsening of sepsis-related outcomes. The kidney, in addition to its role in maintaining body homeostasis, is an endocrine and immune organ capable of generating acute phase and inflammatory response, thereby possibly modulating distant multiorgan dysfunction [26, 37]. Several recent experimental studies have shown that AKI can affect functions of remote organs via multiple biological relevant pathways including transcriptomic changes, oxidative stress, inflammation, leukocyte activation and trafficking, loss of vascular functions, apoptosis, and upregulation of various injury-promoting molecules (reviewed in [26]). This experimental evidence has been corroborated by a series of clinical studies from the PICARD program showing elevated plasma cytokines levels, impaired monocyte cytokine production, oxidative stress, insulin resistance and hyperglycemia in patients with AKI [38–40]. However, to what extent the concept of interorgan crosstalk, which has been described mostly in rodent models of kidney ischemia/reperfusion, is applicable to the septic scenario in humans remains speculative. The first experimental study that evaluated mechanisms through which pre-existing kidney injury amplifies sepsis-induced renal dysfunction and worsens outcome in sepsis has recently been published by Doi et al. [41]. Using a 2-stage 'acute-on-chronic' mouse model of folic acid-induced AKI followed by sepsis with cecal ligation and puncture, the authors demonstrated that these animals developed substantially more severe sepsis and SA-AKI with higher mortality rates than induced by sepsis alone. These data clearly indicate that pre-existing renal dysfunction worsens sepsis by amplifying additional kidney damage and promoting septic shock. Increased capillary permeability, plasma vascular endothelial growth factor, decreased bacteria clearance and splenocyte apoptosis were indentified as potential pathomechanisms associated with the synergistic effects of sepsis and kidney injury. Taken together, AKI can add fuel to the existing sepsis-associated inflammation and immune dysregulation that, coupled with reduced renal clearance, may amplify not only kidney damage itself, but also injury of distant organs. Postulated mechanisms underlying the bidirectional relationship between sepsis and AKI are shown in figure 2.

Conclusion

Sepsis is the leading cause of AKI in critically ill patients. Development of AKI during sepsis is associated with poor outcome and independently predicts increased mortality. Conversely, sepsis frequently complicates the course of nonseptic AKI, and is associated with substantial mortality as well. Basic science and clinical data increasingly suggest there may be a bidirectional and synergistic relationship between sepsis and AKI. Both sepsis and AKI lead to the substantial loss of physiological balance of inflammatory, metabolic pathways and oxidative stress control, as well as to the production of a multitude of injury-promoting molecules. When occurring together, the magnitude of this imbalance seems to be enhanced, resulting in the acceleration of the downward spiral toward high morbidity and mortality for this patient population. Hence, AKI can no longer be considered an innocent bystander, but rather the Achilles' heel of critically ill patients driving a broad, yet not completely understood, spectrum of harmful biological responses capable of affecting functions of remote organs and systems.

Acknowledgement

This work was supported by a research project MSM 0021620819 (replacement of and support to some vital organs).

References

1 Martin GS, Mannino DM, Eaton S, Moss M: The epidemiology of sepsis in the United States from 1979 through 2000. N Engl J Med 2003;348:1546–1554.

2 Parmar A, Langenberg C, Wan L, May CN, Bellomo R, Bagshaw SM: Epidemiology of septic acute kidney injury. Curr Drug Targets 2009;10:1169–1178.

3 Schrier RW, Wang W: Acute renal failure and sepsis. N Engl J Med 2004;351:159–169.

4 Bagshaw SM, George C, Bellomo R, ANZICS Database Management Committee: Changes in the incidence and outcome for early acute kidney injury in a cohort of Australian intensive care units. Crit Care 2007;11:R68.

5 Vincent JL, Sakr Y, Sprung CL, Ranieri VM, Reinhart K, Gerlach H, Moreno R, Carlet J, Le Gall JR, Payen D, Sepsis Occurrence in Acutely Ill Patients Investigators: Sepsis in European intensive care units: results of the SOAP study. Crit Care Med 2006;34:344–353.

6 Ostermann M, Chang RW: Acute kidney injury in the intensive care unit according to RIFLE. Crit Care Med 2007;35:1837–1843.

7 Rangel-Frausto MS, Pittet D, Costigan M, Hwang T, Davis CS, Wenzel RP: The natural history of the systemic inflammatory response syndrome (SIRS). A prospective study. JAMA 1995;273:117–123.

8 Bagshaw SM, Lapinsky S, Dial S, Arabi Y, Dodek P, Wood G, Ellis P, Guzman J, Marshall J, Parrillo JE, Skrobik Y, Kumar A, Cooperative Antimicrobial Therapy of Septic Shock (CATSS) Database Research Group: Acute kidney injury in septic shock: clinical outcomes and impact of duration of hypotension prior to initiation of antimicrobial therapy. Intensive Care Med 2009;35:871–881.

9 Reynvoet E, Vandijck DM, Blot SI, Dhondt AW, De Waele JJ, Claus S, Buyle FM, Vanholder RC, Hoste EA: Epidemiology of infection in critically ill patients with acute renal failure. Crit Care Med 2009;37:2203–2209.

10 Levy EM, Viscoli CM, Horwitz RI: The effect of acute renal failure on mortality. A cohort analysis. JAMA 1996;275:1489–1494.

11 Thakar CV, Yared JP, Worley S, Cotman K, Paganini EP: Renal dysfunction and serious infections after open-heart surgery. Kidney Int 2003;64:239–246.

12 Hoste EA, Blot SI, Lameire NH, Vanholder RC, De Bacquer D, Colardyn FA: Effect of nosocomial bloodstream infection on the outcome of critically ill patients with acute renal failure treated with renal replacement therapy. J Am Soc Nephrol 2004;15:454–462.

13 Mehta RL, Bouchard J, Soroko SB, Ikizler TA, Paganini EP, Chertow GM, Himmelfarb J, Program to Improve Care in Acute Renal Disease (PICARD) Study Group: Sepsis as a cause and consequence of acute kidney injury: Program to Improve Care in Acute Renal Disease. Intensive Care Med 2011;37:241–248.
14 Vandijck DM, Reynvoet E, Blot SI, Vandecasteele E, Hoste EA: Severe infection, sepsis and acute kidney injury. Acta Clin Belg Suppl 2007;2:332–336.
15 Chertow GM, Burdick E, Honour M, Bonventre JV, Bates DW: Acute kidney injury, mortality, length of stay, and costs in hospitalized patients. J Am Soc Nephrol 2005;16:3365–3370.
16 Uchino S, Kellum JA, Bellomo R, Doig GS, Morimatsu H, Morgera S, Schetz M, Tan I, Bouman C, Macedo E, Gibney N, Tolwani A, Ronco C, Beginning and Ending Supportive Therapy for the Kidney (BEST Kidney) Investigators: Acute renal failure in critically ill patients: a multinational, multicenter study. JAMA 2005;294:813–818.
17 Coca SG, Yusuf B, Shlipak MG, Garg AX, Parikh CR: Long-term risk of mortality and other adverse outcomes after acute kidney injury: a systematic review and meta-analysis. Am J Kidney Dis 2009;53:961–973.
18 Bagshaw SM, George C, Bellomo R, ANZICS Database Management Committee: Early acute kidney injury and sepsis: a multicentre evaluation. Crit Care 2008;12:R47.
19 Murugan R, Karajala-Subramanyam V, Lee M, Yende S, Kong L, Carter M, Angus DC, Kellum JA, Genetic and Inflammatory Markers of Sepsis (GenIMS) Investigators: Acute kidney injury in non-severe pneumonia is associated with an increased immune response and lower survival. Kidney Int 2010;77:527–535.
20 Chvojka J, Sýkora R, Karvunidis T, Raděj J, Kroužecký A, Novák I, Matějovič M: New developments in septic acute kidney injury. Physiol Res 2010;59:859–869.
21 Bellomo R, Wan L, Langenberg C, May C: Septic acute kidney injury: new concepts. Nephron Exp Nephrol 2008;109:e95–e100.
22 Lopes JA, Fernandes P, Jorge S, Resina C, Santos C, Pereira A, Neves J, Antunes F, Gomes da Costa A: Long-term risk of mortality after acute kidney injury in patients with sepsis: a contemporary analysis. BMC Nephrol 2010;11:9.
23 Garg AX, Parikh CR: Yin and yang: acute kidney injury and chronic kidney disease. J Am Soc Nephrol 2009;20:8–10.
24 Le Dorze M, Legrand M, Payen D, Ince C: The role of the microcirculation in acute kidney injury. Curr Opin Crit Care 2009;15:503–508.
25 Kinsey GR, Li L, Okusa MD: Inflammation in acute kidney injury. Nephron Exp Nephrol 2008;109:e102–e107.
26 Li X, Hassoun HT, Santora R, Rabb H: Organ crosstalk: the role of the kidney. Curr Opin Crit Care 2009;15:481–487.
27 Chawla LS, Seneff MG, Nelson DR, Williams M, Levy H, Kimmel PL, Macias WL: Elevated plasma concentrations of IL-6 and elevated APACHE II score predict acute kidney injury in patients with severe sepsis. Clin J Am Soc Nephrol 2007;2:22–30.
28 Mariano F, Cantaluppi V, Stella M, Romanazzi GM, Assenzio B, Cairo M, Biancone L, Triolo G, Ranieri VM, Camussi G: Circulating plasma factors induce tubular and glomerular alterations in septic burns patients. Crit Care 2008;12:R42.
29 Sarnak MJ, Jaber BL: Mortality caused by sepsis in patients with end-stage renal disease compared with the general population. Kidney Int 2000;58:1758–1764.
30 Bouchard J, Soroko SB, Chertow GM, Himmelfarb J, Ikizler TA, Paganini EP, Mehta RL, Program to Improve Care in Acute Renal Disease (PICARD) Study Group: Fluid accumulation, survival and recovery of kidney function in critically ill patients with acute kidney injury. Kidney Int 2009;76:422–427.
31 Payen D, de Pont AC, Sakr Y, Spies C, Reinhart K, Vincent JL, Sepsis Occurrence in Acutely Ill Patients (SOAP) Investigators: A positive fluid balance is associated with a worse outcome in patients with acute renal failure. Crit Care 2008;12:R74.
32 Prowle JR, Echeverri JE, Ligabo EV, Ronco C, Bellomo R: Fluid balance and acute kidney injury. Nat Rev Nephrol 2010;6:107–115.

33 Meyer TW, Hostetter TH: Uremia. N Engl J Med 2007;357:1316–1325.
34 Hassoun HT, Grigoryev DN, Lie ML, Liu M, Cheadle C, Tuder RM, Rabb H: Ischemic acute kidney injury induces a distant organ functional and genomic response distinguishable from bilateral nephrectomy. Am J Physiol Renal Physiol 2007;293:F30–F40.
35 Brun-Buisson C, Meshaka P, Pinton P, Vallet B, EPISEPSIS Study Group: EPISEPSIS: a reappraisal of the epidemiology and outcome of severe sepsis in French intensive care units. Intensive Care Med 2004;30:580–588.
36 Vanholder R, De Vriese A, Lameire N: The role of dialyzer biocompatibility in acute renal failure. Blood Purif 2000;18:1–12.
37 Luyckx VA, Cairo LV, Compston CA, Phan WL, Mueller TF: Oncostatin M pathway plays a major role in the renal acute phase response. Am J Physiol Renal Physiol 2009;296:F875–F883.
38 Himmelfarb J, Le P, Klenzak J, Freedman S, McMenamin ME, Ikizler TA, PICARD Group: Impaired monocyte cytokine production in critically ill patients with acute renal failure. Kidney Int 2004;66:2354–2360.
39 Himmelfarb J, McMonagle E, Freedman S, Klenzak J, McMenamin E, Le P, Pupim LB, Ikizler TA, The PICARD Group: Oxidative stress is increased in critically ill patients with acute renal failure. J Am Soc Nephrol 2004;15:2449–2456.
40 Basi S, Pupim LB, Simmons EM, Sezer MT, Shyr Y, Freedman S, Chertow GM, Mehta RL, Paganini E, Himmelfarb J, Ikizler TA: Insulin resistance in critically ill patients with acute renal failure. Am J Physiol Renal Physiol 2005;289:F259–F264.
41 Doi K, Leelahavanichkul A, Hu X, Sidransky KL, Zhou H, Qin Y, Eisner C, Schnermann J, Yuen PS, Star RA: Pre-existing renal disease promotes sepsis-induced acute kidney injury and worsens outcome. Kidney Int 2008;74:1017–1025.

Martin Matejovic, MD, PhD
ICU, 1st Medical Department
Charles University Medical School and Teaching Hospital
alej Svobody 80, 304 60 Plzen (Czech Republic)
Tel. +420 37 710 3501, E-Mail matejovic@fnplzen.cz

Kellum JA, Ronco C, Vincent J-L (eds): Controversies in Acute Kidney Injury.
Contrib Nephrol. Basel, Karger, 2011, vol 174, pp 89–97

Pathophysiology of Sepsis-Induced Acute Kidney Injury: The Role of Global Renal Blood Flow and Renal Vascular Resistance

Adrien Bouglé · Jacques Duranteau

Département d'Anesthésie-Réanimation Chirurgicale, Université Paris XI Sud, Hôpital de Bicêtre, Assistance Publique, Hôpitaux de Paris, Le Kremlin Bicêtre, France

Abstract

Sepsis-induced acute kidney injury (AKI) is the most common form of AKI observed in critically ill patients. AKI mortality in septic critically ill patients remains high despite our increasing ability to support vital organ systems. This high mortality is partly due to our poor understanding of the pathophysiological mechanisms of sepsis-induced AKI. Recent experimental studies have suggested that the pathogenesis of sepsis-induced AKI is much more complex than isolated hypoperfusion due to decreased cardiac output and hypotension. In nonresuscitated septic patients with a low cardiac output, a decrease in renal blood flow (RBF) could contribute to the development of AKI. In resuscitated septic patients with a hyperdynamic circulatory state, RBF is unchanged or increased. However, in resuscitated septic patients, sepsis-induced AKI can occur in the setting of renal hyperemia in the absence of renal hypoperfusion or renal ischemia. Alterations in the microcirculation in the renal cortex or renal medulla can occur despite normal or increased global RBF. Increased renal vascular resistance (RVR) may represent a key hemodynamic factor that is involved in sepsis-associated AKI. Sepsis-induced renal microvascular alterations (vasoconstriction, capillary leak syndrome with tissue edema, leukocytes and platelet adhesion with endothelial dysfunction and/or microthrombosis) and/or an increase in intra-abdominal pressure could contribute to an increase in RVR. Further studies are needed to explore the time course of renal microvascular alterations during sepsis as well as the initiation and development of AKI. Doppler ultrasonography combined with the calculation of the resistive indices may indicate the extent of the vascular resistance changes and may help predict persistent AKI and determine the optimal systemic hemodynamics required for renal perfusion.

Sepsis-induced acute kidney injury (AKI) is the most common form of AKI observed in critically ill patients. AKI mortality in septic and critically ill patients remains high despite our increasing ability to support vital organs. This is partly due to our poor understanding of the pathophysiological mechanisms of sepsis-induced AKI. Recent experimental studies have suggested that the pathogenesis of sepsis-induced AKI is much more complex than an isolated hypoperfusion due to decreased cardiac output and hypotension. This chapter focuses on the recent advances in this area of research and discusses the role of global renal blood flow (RBF) and renal vascular resistance (RVR) in the pathogenesis of sepsis-induced AKI.

Is Sepsis-Induced Acute Kidney Injury Due to Decreased Renal Blood Flow?

The role of RBF alterations in the pathogenesis of sepsis-induced AKI is of critical clinical importance and is still a subject of ongoing debate. RBF values during sepsis are still poorly understood because animal studies have been heterogeneous in their design, have employed a variety of different models to simulate sepsis and have used various techniques to monitor RBF. In addition, the number of human studies is limited because they require renal vein sampling to accurately evaluate RBF.

For many years, it was thought that a decrease in RBF associated with renal vasoconstriction without tubular dysfunction was the predominant pathogenic factor in sepsis-induced AKI [1, 2]. The resulting renal hypoperfusion was thought to be responsible for an ischemic insult due to the diminished oxygen delivery to the tubule cells associated with a sepsis-induced relative increase in oxygen demand. During sepsis, generalized arterial vasodilatation with an associated decrease in systemic vascular resistance induces a decrease in cardiac preload which activates the neurohumoral axis, the sympathetic nervous system, the renin-angiotensin-aldosterone axis and the nonosmotic release of vasopressin (AVP). If this neurohumoral axis activation is essential to maintain the stability of the cardiovascular system in patients with severe sepsis and septic shock, the activation of pressor hormones could lead to renal vasoconstriction, ischemic injuries and AKI. The glomerular filtration rate (GFR) is determined by the net difference in the arterial pressure between the afferent and efferent arterioles across the glomerular capillary bed (the intraglomerular filtration pressure is the physical driving force for GFR). The activation of the sympathetic nervous system with the release of norepinephrine could profoundly constrict the afferent glomerular arterioles, which results in a decrease in the filtration pressure and may thus contribute to AKI in patients with sepsis [1]. The study by Wang et al. [3], published in 2002, illustrated this theory. The authors achieved selective blockade of the renal nerves prior to the injection of lipopolysaccharide (LPS). In this endotoxemic model of normotensive sepsis, the injection of LPS without the blockade of the renal nerves resulted in an increase in the plasma epinephrine and norepinephrine levels that was associated with a decrease in RBF and GFR. In contrast, after renal denervation, no changes in the RBF were observed after LPS injection [3]. Thus, according to these authors, the activation of the neurohumoral axis during sepsis is a crucial factor in sepsis-associated AKI.

However, other experimental studies did not report RBF decreases during sepsis [4–7]. For example, in a porcine model of resuscitated sepsis, Cohen et al. [4] observed that LPS infusion decreased mean arterial pressure (MAP) with an increase in cardiac output (CO), RBF and nitric oxide content in the renal medulla. The increased RBF was diverted to the medulla with no increase in cortical blood flow (medullary flow constituted approximately 45% of total RBF). Selective and nonselective nitric oxide synthase inhibitors increased MAP but decreased RBF. This protective role of endothelial nitric oxide synthase during sepsis-associated AKI has been confirmed by Wang et al. [8].

The different models of sepsis have not elucidated the exact features of RBF during sepsis. The paradigm of sepsis-associated AKI caused by renal vasoconstriction and ischemia was recently challenged by an exhaustive review by Langenberg et al. [5]. The authors reviewed experimental animal and human studies of septic acute renal failure in which RBF was measured. In addition, they assessed the influence of various factors on RBF in the selected studies, including the size of the animal, the technique used for RBF measurements (which included direct measurement via a flow probe, microsphere technique, para-aminohippurate clearance or video microscopy), the level of consciousness of the animals, the recovery period between the preparation surgery and the experiments, the timing of RBF measurement in relation to the septic insult, the method used to induce sepsis, the specifics of fluid administration, CO, and the hemodynamic patterns (hypodynamic and hyperdynamic sepsis). They found that 62% of the animal studies reported a decrease in RBF, whereas the remaining 38% reported no change or an increase in RBF. Using a multivariate logistic regression analysis, they found that only CO was an independent determiner of RBF. Low CO predicted decreased RBF, and preserved or high CO predicted unchanged or increased RBF [5]. Overall, 90% of the studies in which the animals experienced a hypodynamic state showed reduced RBF. However, when a hyperdynamic state was present, which occurs in septic patients after fluid resuscitation, 67% of the studies reported unchanged or increased RBF [5].

In terms of human studies, Langenberg et al. [5] found only three studies that were conducted on septic patients in which RBF was accurately measured using an invasive method, such as a renal vein catheter [9–11]. Despite methodological limitations, these studies suggest that RBF is increased in human sepsis. The conclusions of the Langenberg study [5] were confirmed by two recent studies. In a model of sepsis induced by a continuous administration of *Escherichia coli*, the development of hyperdynamic sepsis was associated with an increase in CO,

edema and increase the renal microvascular resistance which could contribute to microvascular dysfunction. Further studies are needed to explore the time course of renal microvascular alterations during sepsis and the initiation and development of AKI.

Is it possible to evaluate RVR in our clinical practice with renal Doppler ultrasonography? Changes in the Doppler waveform pulsatility are widely interpreted to be indicative of vascular resistance changes. The Doppler resistive index (RI) described by Pourcelot [(peak systolic velocity – end diastolic velocity)/peak systolic velocity] is a traditional waveform index that is used as a measure of vascular resistance. Its value is independent of the diameter of the artery and the angle between the ultrasound beam and blood flow. A value of 0.60 is considered to be a normal value for renal RI, whereas 0.70 is usually considered to be the upper threshold of normal RI in adults [24, 25]. This index has been proposed as a useful parameter for the quantification of the changes in RVR. However, RI does not depend only on the RVR and the relationship between these two RI and RVR can be affected by other potentially confounding factors.

Doppler renal arterial waveform results from the interaction between several factors, particularly vascular resistance, vascular compliance and pulse pressure. Thus, clinicians should consider these factors when examining data from renal Doppler ultrasound in intensive care unit (ICU) settings. The first clinical application of RI in the field of intensive care that showed RI as a predictive factor of AKI development was reported by Lerolle et al. [26]. In 37 septic shock patients, a significant difference in RI obtained at admission was observed between patients with and without renal dysfunction 5 days after inclusion (0.77 vs. 0.68, respectively; $p < 0.001$). RI >0.74 at inclusion had a sensitivity of 78% (95% CI: 52–94), a specificity of 77% (95% CI: 50–93) and a positive likelihood ratio of 3.3 (95% CI: 1.1–35) for severe renal dysfunction 5 days after inclusion.

For the first time, a noninvasive and reproductive tool could successfully predict AKI. After excluding obstruction, the classic diagnostic algorithm of AKI consists of distinguishing prerenal azotemia from acute tubular necrosis (ATN). Prerenal azotemia is assumed to be due to a decrease in the GFR, but is not associated with histopathological renal injury [27]. Despite their current use in daily clinical practice, there is no evidence that urine tests can discriminate prerenal azotemia from ATN, particularly in septic patients [28, 29]. Hence, urinary sodium, fractional excretion of sodium [30] or renal failure index [31] cannot be used after diuretic administration, after hemodialysis or in the presence of anuria. Therefore, ICU physicians need a tool to discriminate prerenal azotemia from ATN. Again, renal Doppler ultrasound may be an attractive tool in this setting.

In 91 patients with AKI, Platt et al. [32] demonstrated that mean RI was significantly higher in patients with ATN (0.85) compared to patients with prerenal azotemia (0.67). In addition, in a population of ICU patients, RI was recently shown to be significantly different between patients without AKI, those with prerenal azotemia and those with ATN (0.71, 0.71 and 0.82, respectively) [33]. Logistic regression analysis revealed that RI >0.795 was predictive of persistent AKI. Moreover, in the Platt et al. study [32], RI was better than urinary indices in the diagnosis of persistent AKI. Finally, renal RI could be useful in the assessment of renal circulation during the initial resuscitation in septic shock.

Even if MAP ≥65 mm Hg is recommended [34], the adequate MAP level for renal perfusion is still under debate [35]. In 11 patients with septic shock, our group showed a significant decrease in renal RI when increasing the MAP with norepinephrine from 65 to 75 mm Hg [36]. The study was not designed to detect any improvement in renal function, but the increase of MAP from 65 to 75 mm Hg was associated with increased diuresis [36]. Hence, Doppler ultrasonography with RI calculation can be used in the management of septic patients to determine the optimal MAP levels required for renal perfusion.

Conclusion

In the early phase of nonresuscitated septic animals and patients with a low CO, a decrease in RBF could contribute to the development of AKI. However, in resuscitated septic animals and patients with a hyperdynamic circulatory state, RBF is unchanged or increased, and sepsis-induced AKI can still occur in the setting of renal hyperemia in the absence of renal hypoperfusion. Alterations in microcirculation in the renal cortex or the renal medulla can occur despite normal or increased global RBF. Increased RVR may represent a key hemodynamic factor that is involved in sepsis-associated AKI. A sepsis-induced capillary leak could promote the development of tissue edema and increase the renal microvascular resistance, which could contribute to microvascular dysfunction. Further studies are needed to explore the time course of renal microvascular alterations during sepsis and the initiation and development of AKI. Doppler ultrasonography with RI calculation can be useful as an indicator of vascular resistance changes, in the prediction of persistent AKI and in the determination of the optimal systemic hemodynamics required for renal perfusion.

References

1 Schrier RW, Wang W: Acute renal failure and sepsis. N Engl J Med 2004;351:159–169.

2 Schrier RW, Wang W, Poole B, Mitra A: Acute renal failure: definitions, diagnosis, pathogenesis, and therapy. J Clin Invest 2004;114:5–14.

3 Wang W, Falk SA, Jittikanont S, Gengaro PE, Edelstein CL, Schrier RW: Protective effect of renal denervation on normotensive endotoxemia-induced acute renal failure in mice. Am J Physiol Renal Physiol 2002;283:F583–F587.

4 Cohen RI, Hassell AM, Marzouk K, Marini C, Liu SF, Scharf SM: Renal effects of nitric oxide in endotoxemia. Am J Respir Crit Care Med 2001;164:1890–1895.
5 Langenberg C, Bellomo R, May C, Wan L, Egi M, Morgera S: Renal blood flow in sepsis. Crit Care 2005;9:R363–R374.
6 Langenberg C, Wan L, Egi M, May CN, Bellomo R: Renal blood flow in experimental septic acute renal failure. Kidney Int 2006;69:1996–2002.
7 Chvojka J, Sykora R, Krouzecky A, Radej J, Varnerova V, Karvunidis T, Hes O, Novak I, Radermacher P, Matejovic M: Renal haemodynamic, microcirculatory, metabolic and histopathological responses to peritonitis-induced septic shock in pigs. Crit Care 2008;12:R164.
8 Wang W, Mitra A, Poole B, Falk S, Lucia MS, Tayal S, Schrier R: Endothelial nitric oxide synthase-deficient mice exhibit increased susceptibility to endotoxin-induced acute renal failure. Am J Physiol Renal Physiol 2004;287:F1044–F1048.
9 Lucas CE, Rector FE, Werner M, Rosenberg IK: Altered renal homeostasis with acute sepsis. Clinical significance. Arch Surg 1973;106:444–449.
10 Rector F, Goyal S, Rosenberg IK, Lucas CE: Sepsis: a mechanism for vasodilatation in the kidney. Ann Surg 1973;178:222–226.
11 Brenner M, Schaer GL, Mallory DL, Suffredini AF, Parrillo JE: Detection of renal blood flow abnormalities in septic and critically ill patients using a newly designed indwelling thermodilution renal vein catheter. Chest 1990;98:170–179.
12 Langenberg C, Wan L, Egi M, May CN, Bellomo R: Renal blood flow and function during recovery from experimental septic acute kidney injury. Intensive Care Med 2007;33:1614–1618.
13 Langenberg C, Bagshaw SM, May CN, Bellomo R: The histopathology of septic acute kidney injury: a systematic review. Crit Care 2008;12:R38.
14 Brodsky SV, Yamamoto T, Tada T, Kim B, Chen J, Kajiya F, Goligorsky MS: Endothelial dysfunction in ischemic acute renal failure: rescue by transplanted endothelial cells. Am J Physiol Renal Physiol 2002;282:F1140–F1149.
15 Yamamoto T, Tada T, Brodsky SV, Tanaka H, Noiri E, Kajiya F, Goligorsky MS: Intravital videomicroscopy of peritubular capillaries in renal ischemia. Am J Physiol Renal Physiol 2002;282:F1150–F1155.
16 Sutton TA, Fisher CJ, Molitoris BA: Microvascular endothelial injury and dysfunction during ischemic acute renal failure. Kidney Int 2002;62:1539–1549.
17 Legrand M, Mik EG, Johannes T, Payen D, Ince C: Renal hypoxia and dysoxia after reperfusion of the ischemic kidney. Mol Med 2008;14:502–516.
18 Saotome T, Ishikawa K, May CN, Birchall IE, Bellomo R: The impact of experimental hypoperfusion on subsequent kidney function. Intensive Care Med;36:533–540.
19 Wan L, Bagshaw SM, Langenberg C, Saotome T, May C, Bellomo R: Pathophysiology of septic acute kidney injury: what do we really know? Crit Care Med 2008;36:S198–S203.
20 Di Giantomasso D, Morimatsu H, May CN, Bellomo R: Intrarenal blood flow distribution in hyperdynamic septic shock: effect of norepinephrine. Crit Care Med 2003;31:2509–2513.
21 Redfors B, Bragadottir G, Sellgren J, Sward K, Ricksten SE: Acute renal failure is NOT an 'acute renal success' – a clinical study on the renal oxygen supply/demand relationship in acute kidney injury. Crit Care Med 2010;38:1695–1701.
22 Horbelt M, Lee SY, Mang HE, Knipe NL, Sado Y, Kribben A, Sutton TA: Acute and chronic microvascular alterations in a mouse model of ischemic acute kidney injury. Am J Physiol Renal Physiol 2007;293:F688–F695.
23 Sutton TA: Alteration of microvascular permeability in acute kidney injury. Microvasc Res 2009;77:4–7.
24 Tublin ME, Bude RO, Platt JF: Review. The resistive index in renal Doppler sonography: where do we stand? AJR Am J Roentgenol 2003;180:885–892.
25 Keogan MT, Kliewer MA, Hertzberg BS, DeLong DM, Tupler RH, Carroll BA: Renal resistive indexes: variability in Doppler US measurement in a healthy population. Radiology 1996;199:165–169.

26 Lerolle N, Guerot E, Faisy C, Bornstain C, Diehl JL, Fagon JY: Renal failure in septic shock: predictive value of Doppler-based renal arterial resistive index. Intensive Care Med 2006;32:1553–1559.
27 Lameire N, Van Biesen W, Vanholder R: Acute renal failure. Lancet 2005;365:417–430.
28 Bagshaw SM, Langenberg C, Haase M, Wan L, May CN, Bellomo R: Urinary biomarkers in septic acute kidney injury. Intensive Care Med 2007;33:1285–1296.
29 Bellomo R, Bagshaw S, Langenberg C, Ronco C: Pre-renal azotemia: a flawed paradigm in critically ill septic patients?; in Ronco C, Bellomo R, Kellum JA (eds): Acute Kidney Injury. Contrib Nephrol. Basel, Karger, 2007, vol 156, pp 1–9.
30 Espinel CH: The FENa test. Use in the differential diagnosis of acute renal failure. JAMA 1976;236:579–581.
31 Miller TR, Anderson RJ, Linas SL, Henrich WL, Berns AS, Gabow PA, Schrier RW: Urinary diagnostic indices in acute renal failure: a prospective study. Ann Intern Med 1978;89:47–50.
32 Platt JF, Rubin JM, Ellis JH: Acute renal failure: possible role of duplex Doppler US in distinction between acute prerenal failure and acute tubular necrosis. Radiology 1991;179:419–423.
33 Darmon M, Schortgen F, Vargas F, Liazydi A, Schlemmer B, Brun-Buisson C, Brochard L: Diagnostic accuracy of Doppler renal resistive index for reversibility of acute kidney injury in critically ill patients. Intensive Care Med 2011;37:68–76.
34 Dellinger RP, Carlet JM, Masur H, et al: Surviving Sepsis Campaign guidelines for management of severe sepsis and septic shock. Crit Care Med 2004;32:858–873.
35 Bellomo R, Wan L, May C: Vasoactive drugs and acute kidney injury. Crit Care Med 2008;36:S179–S186.
36 Deruddre S, Cheisson G, Mazoit JX, Vicaut E, Benhamou D, Duranteau J: Renal arterial resistance in septic shock: effects of increasing MAP with norepinephrine on the renal resistive index assessed with Doppler ultrasonography. Intensive Care Med 2007;33:1557–1562.

Jacques Duranteau
Département d'Anesthésie-Réanimation Chirurgicale, EA 3509
Université Paris XI Sud, Hôpital de Bicêtre, Assistance Publique – Hôpitaux de Paris
78 rue du Général Leclerc, FR–94275 Le Kremlin-Bicêtre Cedex (France)
Tel. +33 1 4521 3447, E-Mail jacques.duranteau@bct.aphp.fr

Kellum JA, Ronco C, Vincent J-L (eds): Controversies in Acute Kidney Injury.
Contrib Nephrol. Basel, Karger, 2011, vol 174, pp 98–107

Septic Acute Kidney Injury: The Glomerular Arterioles

Rinaldo Bellomo · Li Wan · Christoph Langenberg · Ken Ishikawa · Clive N. May

Department of Intensive Care and Department of Medicine, Austin Health, and Howard Florey Institute, University of Melbourne, Melbourne, Australia

Abstract

Acute kidney injury (AKI) is a serious condition that affects many intensive care unit (ICU) patients. The most common causes of AKI in the ICU are severe sepsis and septic shock. The mortality of AKI in septic critically ill patients remains high despite our increasing ability to support vital organs. This is partly due to our poor understanding of the pathogenesis of sepsis-induced renal dysfunction. However, new concepts are emerging to explain the pathogenesis of septic AKI, which challenge previously held dogma. Throughout the past half century, septic AKI has essentially been considered secondary to tubular injury, which, in turn, has been considered secondary to renal ischemia. This belief is curious because the hallmark of septic AKI and AKI in general is the loss of glomerular filtration rate (GFR). It would seem logical, therefore, to focus on the glomerulus in trying to understand why such loss of GFR occurs. Recent experimental observations suggest that, at least in the initial phases of septic AKI, profound changes occur which involve glomerular hemodynamics and lead to loss of GFR. These observations imply that changes in the vasoconstrictor tone of both the afferent and efferent arterioles are an important component of the pathogenesis of septic AKI.

Severe acute kidney injury (AKI) requiring acute renal replacement therapy affects approximately 5% of all intensive care unit (ICU) patients [1]. When recently published consensus criteria (RIFLE criteria) are used to define AKI [2] this condition can be shown to occur in almost 8% of hospital patients [3] and in >50% of ICU patients [4]. Thus, AKI is a major clinical problem in modern hospitals and ICUs. In addition, there is strong evidence that sepsis and septic shock are the most important cause of AKI in critically ill patients and account for 50% or more of cases of AKI in the ICU [1, 5]. Despite our increasing ability to support vital organs and resuscitate patients, the incidence and mortality of septic AKI remain high [1, 5, 6]. A possible explanation of why, despite treatment, AKI is so common in severe sepsis and septic shock and why mortality has remained high might relate to our limited understanding of septic AKI and its pathogenesis.

The Pathogenesis of Septic Acute Kidney Injury

Our understanding of the pathogenesis of human AKI in general is markedly affected by the lack of histopathological information and the limited value of blood tests and urine tests in providing us with meaningful information. It is not surprising, therefore, that our understanding of septic human AKI has advanced little in the last 50 years. To overcome such limitations, animal models of AKI have been developed that enable more sophisticated and invasive measurements to be made. Unfortunately, as recently highlighted [7], these animal models have been mostly based on ischemia-reperfusion injury or drug-induced injury. These models are not relevant to septic AKI.

The Primacy of Renal Blood Flow

A major paradigm that has been derived from observations in animals and humans with hypodynamic shock (hemorrhagic, cardiogenic or even septic) is that AKI is due to renal ischemia. This construct implies that restoration of adequate renal blood flow should therefore be the primary means of renal protection in critically ill patients. However, whether in the more commonly seen hyperdynamic septic AKI, renal blood flow (RBF) actually decreases significantly, remains stable or even increases, remains controversial and difficult to study. Accordingly, the dogma that most AKI (including septic AKI) is 'ischemic' remains inadequately tested.

In several experimental studies of septic acute renal failure, global RBF declines after induction of sepsis or endotoxemia [8, 9]. This may result not only in a reduction in glomerular filtration rate (GFR), but also in metabolic deterioration and diminished contents of high-energy phosphates if hypoperfusion is severe and prolonged, possibly causing cell death, acute tubular necrosis and established acute renal failure.

On the other hand, other studies have found that the renal circulation participates in the systemic vasodilatation seen during severe sepsis/septic shock, so that RBF does not diminish, but septic AKI develops. Thus AKI, at least in the first 6–48 h, may develop not in the setting of renal hypoperfusion, but in the setting of maintained and even increased renal perfusion.

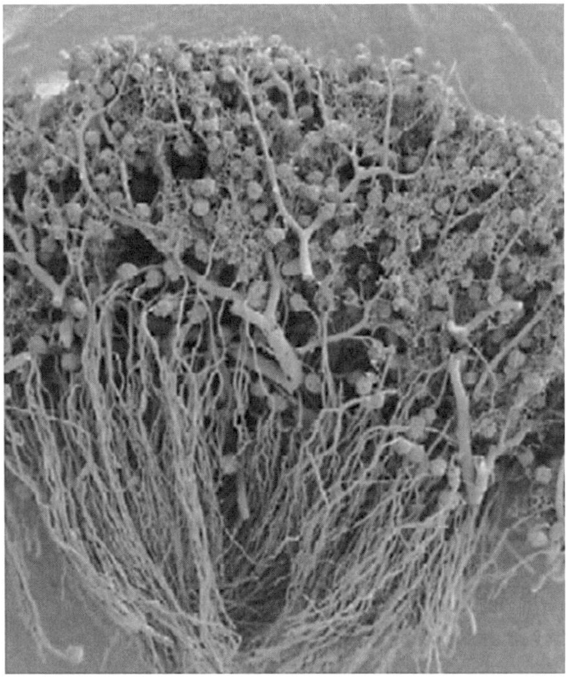

Fig. 2. Electron scanning microscope view of a renal circulation cast showing how the glomeruli represent the final blood flow delivery site for almost all of the renal circulation.

Sustained Septic Acute Kidney Injury and the Glomerular Circulation

The above observations, while important, only related to a short-term model of septic AKI (6 h). It was unknown whether RBF remained elevated if hyperdynamic sepsis was sustained for a longer period. In order to address this issue of relevance to human sepsis, we developed a longer-term model (48 h) of septic AKI achieved by the continuous intravenous infusion of live *E. coli* at a dose carrying limited lethality (20%). In this model, we were able to maintain a high cardiac output, low blood pressure state for 48 h [20]. We found that cardiac output increased threefold over time and that RBF increased at almost exactly the same rate as a mirror image of the cardiac output. This increase in RBF could be fully accounted for by renal vasodilatation. As RBF increased by 300%, however, urinary output progressively declined to near anuria and creatinine clearance (our marker for GFR) decreased by 80%. The serum creatinine increased fourfold. Clearly, these animals had developed severe septic AKI with distinct loss of GFR in the setting of pronounced hyperemia. Interestingly, during this 48-hour period, all markers of tubular function available to us (urinary sodium excretion, fractional excretion of sodium and fractional urea excretion)

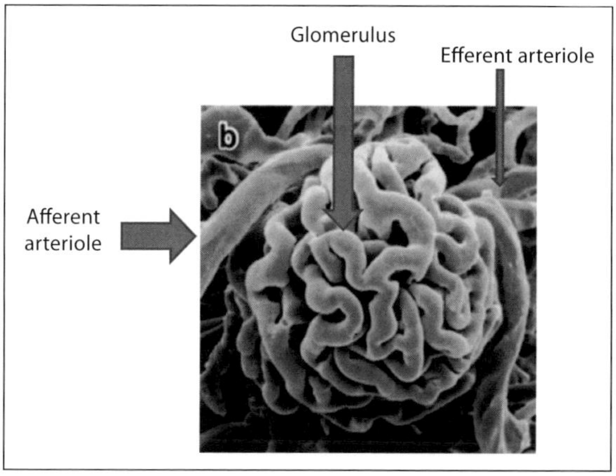

Fig. 3. Scanning electron microscope view of the afferent and efferent arterioles.

indicated preservation of tubular function providing further evidence that (as would be logical) GFR was lost not because of tubular events, but because of glomerular events [20].

We then proceeded to study the recovery from such severe septic AKI. During recovery, we saw the same changes seen during sepsis, but in reverse [21]. RBF returned to normal over 48 h as did renal vascular resistance. The information obtained from this model and the information obtained from the short-term model of septic AKI provided clear proof of concept that oliguric AKI can develop in the presence of marked hyperemia and renal vasodilatation.

These observations call for some reflection of how the mammalian kidney could experience increased RBF, vasodilatation and simultaneous loss of GFR. First, GFR is dependent upon the pressure that drives ultrafiltration across the glomerular membrane [the glomerular capillary pressure (Gcp)]. This pressure is affected by oncotic pressure, and, more markedly, by the relationship between the resistance of the afferent and efferent glomerular arterioles, the vessels which control inflow and outflow from the glomerulus (fig. 3). These arterioles are also the major controllers of global RBF because >90% of such flow is delivered to the glomeruli. For RBF to increase and renal vascular resistance to decrease, both the efferent and afferent arterioles must dilate. Although such dilatation can account for the hyperemia and renal vasodilatation, it cannot explain the loss of GFR. If the afferent arteriole dilated, with no change in the efferent arteriole, then GFR might be expected to increase because the Gcp would inevitably increase. Thus, for Gcp to decrease, the efferent arterioles must also dilate. However, this is still insufficient to explain what we saw in our experimental animals. If such postulated efferent arteriolar vasodilatation simply maintained the normal relationship between afferent and efferent arteriolar tone, Gcp would be increased by the increased flow and GRF would increase. There is, therefore, a

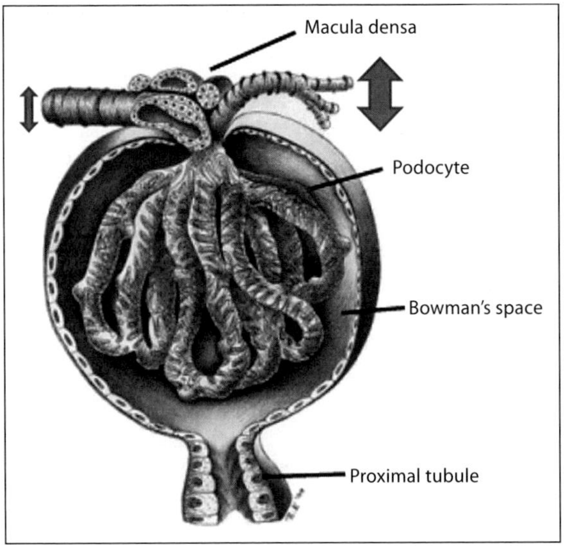

Fig. 4. Figure illustrating likely changes in vessel tone during severe sepsis? The efferent arteriole dilates more than the afferent arteriole, thus changing intraglomerular filtration pressure and decreasing ultrafiltration.

physiologically logical mechanism by which RBF can increase, renal vasodilatation can occur and, simultaneously, Gcp and GFR can decrease – simultaneous dilatation of both arterioles, but with greater efferent than afferent dilatation (fig. 4). This is a physiological state similar to that seen with the administration of angiotensin-converting enzyme inhibitors and characterized by afferent vasodilatation (mild to moderate) and even greater (marked) efferent arteriolar vasodilatation.

If this physiological paradigm were true, then administration of an agent which acts as a preferential efferent arteriolar vasoconstrictor should deliver the following physiological consequences: decreased RBF, renal vasoconstriction, and increased urinary output and GFR. Such an agent exists: angiotensin II. Accordingly, we proceeded to perform controlled studies in our animals to compare intravenous infusion of angiotensin II with placebo in the setting of experimental septic AKI [22]. These studies offered further support for our hypothesis: urinary output increased dramatically as did creatinine clearance (fig. 5), while RBF decreased and renal vasoconstriction developed.

Other studies support the view that preferential efferent arteriolar vasoconstriction might improve GFR. Some evidence exists, for example, that arginine vasopressin (AVP) causes calcium release from intracellular stores, which in turn contributes to the contractile response of both afferent and efferent arteriole [23]. However, such response is more prominent in the efferent arteriole [23] and appears to lead to increases in Gcp [24]. If this paradigm were correct, one

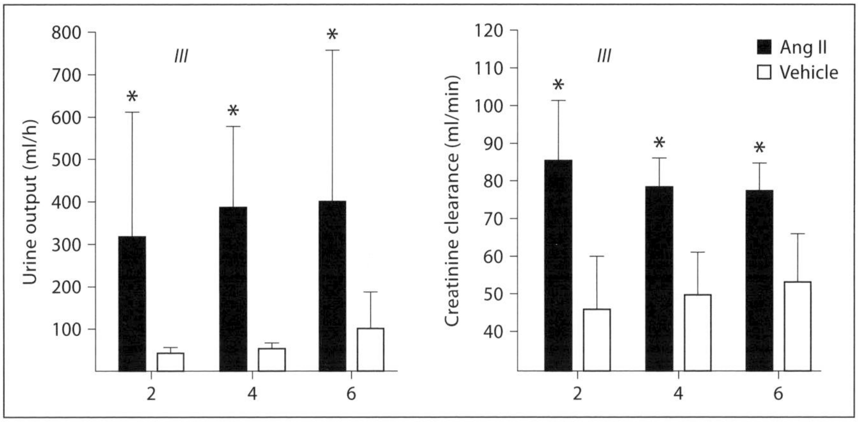

Fig. 5. Changes in urine output and creatinine clearance induced in experimental animals with intravenous infusion of angiotensin II (Ang II). Administration of this preferential efferent arteriolar vasoconstrictor was associated with a major improvement in urine output and creatinine clearance. This change occurred while renal vascular resistance increased and RBF decreased (data not shown – see [22]).

might expect that the infusion of AVP would improve urine output and creatinine clearance in sepsis. Indeed, this is what we were able to show in our model of septic AKI [25]. More importantly, this is what was shown by the VASST investigators in their recent post-hoc analysis of the evolution of AKI in the 464 patients that were part of their multicenter double-blind randomized controlled trial of AVP in septic shock [26]. In patients with RIFLE class R, AVP infusion was associated with a decreased progression to failure and a lower risk of use of renal replacement therapy (17 vs. 37.7%; $p = 0.02$). In their aggregate, these observations continue to provide support for the concept that changes in the relationship between afferent and efferent arteriolar vasoconstrictor tone might well continue to regulate GFR even in patients with developing septic AKI.

Conclusions

As discussed above, a growing body of experimental data has accumulated over the last 5 years to demonstrate that, at least in its initial phases, septic AKI may occur in the setting of marked hyperemia and vasodilatation, and that renal ischemia is not necessary for the loss of GFR to occur in sepsis. Furthermore, evidence is accumulating that changes in the relation of afferent-efferent arteriolar resistance might modulate GFR in both experimental and human septic AKI.

Many questions remain unanswered, however, regarding the mechanisms responsible for such putative preferential efferent arteriolar vasodilatation, its duration, and whether it represents only a short-lived event or whether it continues over time during AKI. We also do not know whether the loss of GFR seen under such circumstances is due to these vascular changes alone, whether backleak of filtered solutes is partly responsible for solute accumulation, or whether a degree of shunting (blood does not enter the glomerulus but actually transits around it and enters the vasa recta in a postglomerular site) occurs through components of the periglomerular circulation that are not yet fully understood. In addition, we do not know how tubular injury and glumerular hemodynamics influence each other in causing and maintaining loss of function is AKI. In our opinion, the evidence available, together with these major uncertainties, justify the continued pursuit of a research agenda into glomerular and periglomerular hemodynamics in septic AKI

References

1 Uchino S, Kellum J, Bellomo R, et al: Acute renal failure in critically ill patients – a multinational, multicenter study. JAMA 2005;294:813–818.

2 Bellomo R, Ronco C, Kellum JA, Mehta RL, Palevsky P: Acute Dialysis Quality Initiative Workgroup: Acute renal failure-definition, outcome measures, animal models, fluid therapy and information technology needs: the second international consensus conference of the ADQI Group. Crit Care 2004;8:R204–R212.

3 Uchino S, Bellomo R, Goldsmith D, Bates S, Ronco C: An assessment of the RIFLE criteria for acute renal failure in hospitalized patients. Crit Care Med 2006;34:1913–1917.

4 Hoste EA, Clermont G, Kersten A, et al: RIFLE criteria for acute kidney injury are associated with hospital mortality in critically ill patients: a cohort analysis. Crti Care 2006;10:R73–R80.

5 Silvester W, Bellomo R, Cole L: Epidemiology, management, and outcome of severe acute renal failure of critical illness in Australia. Crit Care Med 2001;29:1910–1915.

6 Bagshaw SM, Uchino S, Bellomo R, et al: Septic acute kidney injury in critically ill patients: clinical characteristics and outcome. Clin J Am Soc Nephrol 2007;2:431–439.

7 Heyman SN, et al: Animal models of acute tubular necrosis. Curr Opin Crit Care 2002;8:526–534.

8 Badr KF, Kelley VE, Rennke HG, Brenner BM: Roles for thromboxane A_2 and leukotrienes in endotoxin-induced acute renal failure. Kidney Int 1986;30:474–480.

9 Kikkeri D, Pennell JP, Hwang KH, Jacob AI, Richman AV, Bourgoignie JJ: Endotoxemic acute renal failure in awake rats. Am J Physiol 1986;250:F1098–F1106.

10 Ravikant T, Lucas TE: Renal blood flow distribution in septic hyperdynamic pigs. J Surg Res 1977;22:294–298.

11 Brenner M, Schaer GL, Mallory DL, Suffredini AF, Parrillo JE: Detection of renal blood flow abnormalities in septic and critically ill patients using a newly designed indwelling thermodilution renal vein catheter. Chest 1990;98:170–179.

12 Langenberg C, Bellomo R, May C, Wan L, Egi M, Morgera S: Renal blood flow in sepsis. Crit Care 2005;9:R363–R374.

13 Di Giantomasso D, May C, Bellomo R: Vital organ blood flow in hyperdynamic sepsis. Chest 2003;124:1053–1059.

14 Di Giantomasso D, May C, Bellomo R: Norepinephrine and organ blood flow in sepsis. Intensive Care Med 2003;29:1774–1781.

15 Wan L, Bellomo R, May C: The effect of normal saline resuscitation on vital organ blood flow in septic sheep. Intensive Care Med 2006;32:1238–1242.

16 May C, Wan L, Williams J, Wellard MR, Pell G, Jackson G, Bellomo R: A technique for the measurements of renal ATP in a large animal model of sepsis. In J Artif Organs 2005;28:16–21.
17 May C, Wan L, Williams J, Wellard MR, Pell G, Langenberg C, Bellomo R: A technique for the simultaneous measurement of renal ATP, blood flow and pH in a large animal model of septic shock. Crit Care Resusc 2007;9:30–33.
18 DiGiantomasso D, Bellomo R, May CN: The haemodynamic and metabolic effects of epinephrine in experimental hyperdynamic septic shock. Intensive Care Med 2005;31:454–462.
19 Di Giantomasso D, Morimatsu H, May CN, Bellomo R: Intra-renal blood flow distribution in hyperdynamic septic shock: effect of norepinephrine. Crit Care Med 2003;31:2509–2513.
20 Langenberg C, Wan L, Egi M, May CN, Bellomo R: Renal blood flow in experimental septic acute renal failure. Kidney Int 2006;69:1996–2002.
21 Langenberg C, Wan L, Egi M, May CN, Bellomo R: Renal blood flow and function during recovery from experimental septic acute kidney injury. Intensive Care Med 2007;33:1614–1618.
22 Wan L, Langenberg C, Bellomo R, May CN: Angiotensin II in experimental hyperdynamic sepsis. Crit Care 2009:13:R190.
23 Fallett RW, Ikenaga H, Bast JP, Carmines PK: Relative contribution of Ca2+ mobilization and influx in renal arteriolar contractile responses to arginine vasopressin. Am J Physiol Renal Physiol 2005;288: F545–F551.
24 Yamada K, Nakano H, Nishimura M, Yoshida S: Effect of AVP:V1-receptor antagonist on urinary albumin excretion and renal hemodynamics in NIDDM nephropathy: role of AVP-V1 receptor. J Diabetes Complications 1995;9:326–329.
25 Di Giantomasso, Morimatsu H, Bellomo R, May CN: Effect of low dose vasopressin infusion on vital organ blood flow in the conscious normal and septic sheep. Anaesth Intensive Care 2006;34:427–433.
26 Gordon AC, Russell JA, Walley KR, et al: The effect of vasopressin on acute kidney injury in septic shock. Intensive Care Med 2010;36:83–91.

Prof. Rinaldo Bellomo
Department of Intensive Care
Austin Health
Heidelberg, Victoria 3084 (Australia)
Tel. +61 3 9496 5992, E-Mail rinaldo.bellomo@austin.org.au

Kellum JA, Ronco C, Vincent J-L (eds): Controversies in Acute Kidney Injury.
Contrib Nephrol. Basel, Karger, 2011, vol 174, pp 108–118

Kidney Endothelial Dysfunction: Ischemia, Localized Infections and Sepsis

Bruce A. Molitoris · Ruben M. Sandoval

Indiana University Department of Medicine and the Indiana Center for Biological Microscopy, Indianapolis, Ind., USA

Abstract

Endothelial cells play a key role in initiating and propagating the inflammatory response seen in ischemia, infections and sepsis. Situated in a key position between the epithelial cells and white blood cells (WBC), they interact and respond to signals from both cell types. Microvascular endothelial cells within the kidney mediate coagulation, WBC attachment, WBC migration into the interstitium, microvascular flow rates and permeability. Low regeneration potential and endothelial-mesenchymal transformation lead to fibrosis and subsequent microvascular dropout. This last event is in large part responsible for a chronic reduction in regional perfusion, subsequent increased vulnerability to recurrent acute kidney injury, and acceleration of chronic kidney disease progression to end-stage renal disease. Glomerular endothelial dysfunction may lead to preglomerular shunting of blood flow allowing kidney blood flow to remain close to normal while resulting in a reduction in glomerular filtration rate.

Under physiologic conditions, the cells within the kidney function in an integrated, synergistic and 4-D fashion to perform a number of finely tuned and interactive processes resulting in a complexity of cell-cell interactions and tissue heterogeneity second only to the brain. The kidney has one of the richest and most diversified endothelial populations found within any organ. While receiving approximately 20–25% of the cardiac output, the renal endothelium experiences unparalleled environmental extremes in oxygenation and osmolality. For example, endothelial cells in the outer cortex are exposed to a normal osmolality and oxygen tension where those in the inner medullary region are exposed to an osmolality of up to 1,200 mOsm and an oxygenation content down to approximately 20–30 mm Hg. Glomerular endothelial cells are fenestrated as are certain peritubular microvascular endothelial cells.

The kidney endothelial cell is positioned in a strategic position either mediating filtration across the glomerulus or in direct contact with epithelial cells, whose apical membrane face the external milieu, and white blood cells (WBC) that monitor the internal milieu. Crosstalk between these three cell types occurs across and through endothelial cells, and all three cell types, as well as resident dendritic cells, are involved with the inflammatory response within the tissue. Endothelial cells also regulate the coagulation cascade as part of the response to injury and inflammation.

When blood flow diminishes in the kidney, regardless of whether ischemia occurs due to a macrovascular systemic or a microvascular focal event, the result remains the same. Adaptation to a reduction in blood flow can only occur to a certain point and then, when delivery of oxygen and metabolic substrates is inadequate, cellular ATP drops and cellular injury results leading to organ dysfunction. The purpose of this review is to identify, organize and define interactions between the different types of cells and processes involved in the pathophysiology of endothelial dysfunction and acute kidney injury (AKI); detail how bacterial infections mediate endothelial changes through proximal tubule signals; delineate how preglomerular vascular shunting can in part explain near normal total renal blood flow and yet a reduced glomerular filtration rate (GFR); and outline an approach that facilitates patient care and development of therapies.

Lessons Learned from Isolated Tubular Bacterial Infections

In a novel series of intravital 2-photon studies, the laboratory of Agneta Richter-Dahlfors demonstrated how *Escherichia coli* infections isolated to one proximal tubule lumen could dramatically alter endothelial function in a site-specific manner [1–3]. Utilizing a fluorescent cytopathologic strain of *E. coli* and micropuncture delivery to the early proximal tubule (PT), these investigators were able to record in four dimensions (3-D plus time) bacterial attachment to PT cell apical membranes and the resulting local responses in peritubular blood flow and WBC recruitment. Rapidly, within 1–2 h following PT intralumenal bacterial injection and attachment to PT cells, and before there was disruption of the epithelial lining at 5–8 h, there was a marked reduction in peritubular red blood cell flow rates surrounding only the infected tubule. Neutrophils started accumulating at 2–3 h and continued to grow in number, and endothelial permeability increased. These responses were markedly delayed when an identical strain of *E. coli* lacking only one virulence factor was used [1]. Utilizing Clark-type microelectrodes within the lumen of infected tubules in vivo, oxygen tension (PO_2) decreased within 1 h of bacterial injection and was 0 mm Hg within 3.5–4.0 h [3]. This decrease in O_2 tension was not due to bacterial O_2 utilization.

Primary cultures of PT cells showed increased oxygen consumption and release of proinflammatory cytokines including TNF-α, IL-1β and IL-6 when exposed to the bacteria. These same cytokines were increased post-PT lumen injection of bacteria into kidney tissue by PCR techniques 5 h later. These authors also noted an important role for coagulation in the process of bacterial isolation. If infected rats were treated with heparin, there was a marked delay in red blood cell flow rate deterioration surrounding the infected PT with subsequent sepsis and widespread dissemination of bacteria throughout the body. This consequence was not encountered in nonheparin-treated infected rats. These data imply there is a rapid signal sent from the PT cell once bacterial attachment to the apical membrane occurs, resulting in endothelial cell alterations leading to activation of the clotting cascade, WBC recruitment and endothelial dysfunction. This crosstalk has important implications regarding uptake and processing of bacterial products, such as lipopolysaccharide, by PT cells [4]. These data imply that injury to PT cells leads to localized effects on the microcirculation. A general scheme for PT, WBC and endothelial cell interaction is shown in figure 1.

Endothelial Dysfunction

Endothelial cells contribute to vascular tone, regulation of blood flow to local tissue beds, modulation of coagulation and inflammation, and vascular permeability. Both ischemia and sepsis have profound direct and indirect effects on the renal endothelium, resulting in microvascular dysfunction leading to continued ischemic conditions and further injury following the initial insult, especially in the outer stripe of the kidney [5, 6]. Histopathologically, this is seen as vascular congestion, edema formation, diminished microvascular blood flow, and margination and adherence of inflammatory cells to endothelial cells, leading to the extension phase of ischemic AKI [5]. It is important to realize that while a marked decrease in total kidney perfusion results in global ischemia, a fall in the regional perfusion can result in continuing ischemic injury to this region with minimal effects on global perfusion. However, the complexity in vascular beds within the kidney makes interpretation of total kidney blood flow challenging following ischemic or septic injury.

Vascular Tone

Conger et al. [7, 8] were among the first to demonstrate that postischemic rat kidneys cannot autoregulate blood flow even when total renal blood flow had returned to baseline values up to 1 week after injury. Goligorsky et al. [9] found that the constrictor response could be blocked by Ca^{2+} antagonists, and demonstrated loss of normal endothelial nitric oxide synthase (eNOS) function was due to a loss of vasodilator responses to acetylcholine and bradykinin.

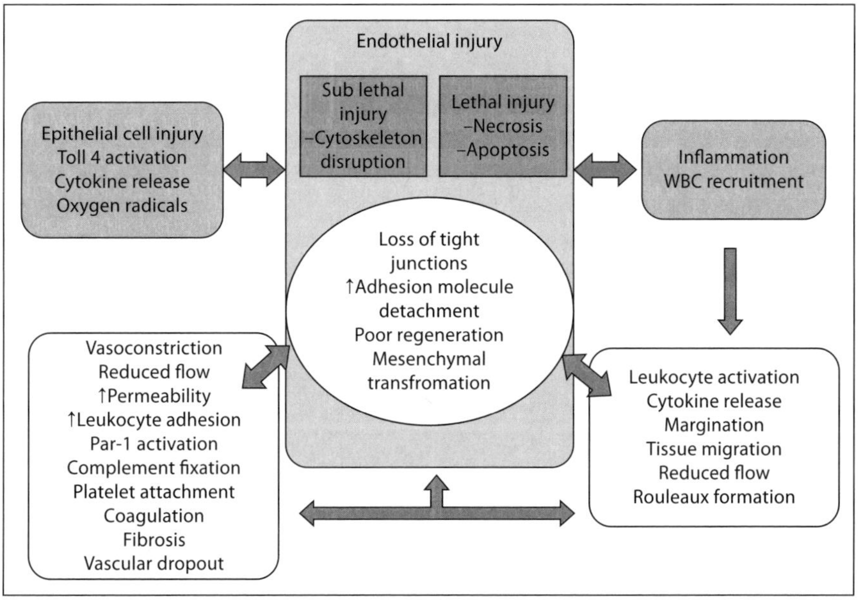

Fig. 1. Events in endothelial cell activation and injury. Ischemia, infection or sepsis cause upregulation and expression of genes encoding for various cell surface proteins such as E- and P-selectin, vascular cell adhesion molecule-1, intercellular adhesion molecule-1 and reduced thrombomodulin. Activated leukocytes adhere to endothelial cells through these adhesion molecules. Endothelial injury increases the production of endothelin-1 and decreases eNOS which serve to induce vasoconstriction and platelet aggregation leading to a hypercoagable state in the microvasculature. The combination of leukocyte adhesion and activation, platelet aggregation, and endothelial injury serves as the platform for vascular congestion of the kidney microvasculature. Defects increase permeability between endothelial cells as a result of tight and adherens junctional alterations. The close proximity and crosstalk of the epithelial PT cells with the microvascular endothelial cells, via release of cytokines and chemokines, further enhance ongoing inflammation. Dendritic cells are also thought to play a major role in this inflammatory cascade and serve as a mediator between the endothelial and epithelial cells.

Selective inhibition, depletion or deletion of inducible NOS have clearly shown renoprotective effects during ischemia [10, 11]. The relative decrease in eNOS, secondary to endothelial dysfunction and damage, may result in a loss of antithrombogenic properties of the endothelium, leading to increased susceptibility to microvascular thrombosis [9]. Administration of L-arginine, NO-donor molsidomine or the eNOS cofactor tetrahydrobiopterin can preserve medullary perfusion and attenuate AKI induced by ischemia/reperfusion (I/R). Conversely, the administration of *N*-nitro-L-arginine methyl ester, an NO blocker, has been reported to aggravate the course of AKI following I/R injury. Although clearly important, these pharmacological studies continue to assess the contribution of eNOS impairment in the overall course of reduced renal function following I/R

[12, 13]. Hence, in ischemic AKI, there is an imbalance of eNOS and inducible NOS leading to tissue injury likely by several different pathways.

Permeability
The endothelial barrier serves to separate the inner space of the blood vessel from the surrounding tissue and to control the exchange of cells and substances between the two. This barrier is diminished during ischemia and sepsis, leading to a loss of integrity [14]. Both transcellular and paracellular pathways participate, the latter being a major contributor to the inflammation-induced barrier dysfunction. Sutton et al. [6, 15] have studied the role of endothelial cells in AKI in a series of experiments utilizing florescent dextrans and 2-photon intravital imaging. The increased microvascular permeability observed in AKI is likely a combination of several factors including loss of endothelial monolayer integrity, alterations in the actin cytoskeleton, breakdown of perivascular matrix, alterations of endothelial cell contacts, upregulated leukocyte-endothelial interactions and severe alterations in the integrity of the adherens junctions of the renal microvasculature [6]. In vivo 2-photon imaging demonstrated a loss of capillary barrier function as early as 4 h of reperfusion with maximal effects seen 24 h after injury.

The breakdown of barrier function may also be due to matrix metalloproteinase-2 or -9 activation, and this upregulation is temporally correlated with an increase in microvascular permeability [5, 15]. In addition, minocycline, a broad-based matrix metalloproteinase inhibitor, and the gelatinase-specific inhibitor ABT-518 both ameliorated the increase in microvascular permeability in this model.

Coagulation
Endothelial cells play a central role in coagulation regulation via interaction with protein C through the endothelial cell protein C receptor and thrombomodulin. Protein C is activated by thrombin-mediated cleavage and the rate of this reaction is further augmented 1,000-fold when thrombin binds to the endothelial cell-surface receptor protein thrombomodulin. The activation rate of protein C is further increased approximately 10-fold when the endothelial cell protein C receptor binds protein C and presents it to the thrombin:thrombomodulin complex. Then, activated protein C essentially has antithrombotic actions, profibrinolytic properties and participates in numerous anti-inflammatory and cytoprotective pathways to restore normal homeostasis [16]. It has also been shown to be an agonist of protease activated receptor-1 [17].

During an inflammatory response, many of the natural anticoagulants including protein C are consumed along with downregulation of the endothelial cell protein C receptor and thrombomodulin expression, which decreases the anticoagulant and anti-inflammatory effects of the protein C pathway. Damaged endothelial cells undergo apoptosis and this further contributes to amplifying the coagulation cascade because the disrupted endothelium provides a procoagulant surface [18]. There is also expression of the tissue factor and von Willebrand factor, as well as interaction with platelets [19]. Activation of the inflammation and coagulation pathways continue to cycle with little control, leading to enhanced microvascular coagulation and further endothelial cell dysfunction. Ultimately, microvascular function is compromised resulting in disseminated intravascular coagulation and microvascular thrombosis, decreased local tissue perfusion and organ dysfunction/failure. It has shown that both pretreatment and postinjury treatment with soluble thrombomodulin attenuates renal injury by minimizing vascular permeabilility defects with improvement in capillary renal blood flow [20].

Both leukocytes and endothelial cells are dynamically involved in the process of adherence of leukocytes to the vascular endothelium. Leukocyte activation including expression of cell surface adhesion molecules and release of cytokines require signals through chemokines circulating in the bloodstream, or through direct contact with the endothelium.

Rolling leukocytes can be activated by chemoattractants such as complement C5a and platelet-activating factor. Once activated, leukocyte integrins bind to endothelial ligands to promote firm adhesion, with β_2-integerin (CD18) being most important. These interactions with the endothelium are mediated through endothelial adhesion molecules that are upregulated during ischemic conditions [21]. Singbartl and Ley [21] have found that platelet P-selectin and not endothelial P-selectin was the main determinant in neutrophil-mediated ischemic renal injury. There is also significant protection from both ischemic injury and mortality by blockade of the shared ligand to all three selectins (E-, P- and L-selectin) which seems to be dependent on the presence of a key fucosyl sugar on the selectin ligand [22, 23]. In a CLP model of septic azotemia, mice gene-deficient for E-selectin or P-selectin, or both, were completely protective.

Long-Term Effects of Endothelial Cell Injury

Recent evidence indicates injury to endothelial cells may have long-term chronic disease implications. Basile [24] documented significant reductions in blood-vessel density following acute ischemic injury leading to the phenomenon of vascular dropout. This was verified by Horbelt et al. [25] who found a drop in the vascular density of almost 45% at 4 weeks after ischemic insult. Endothelial cells undergo mesenchymal tramsformation and lack regenerative potential [26]. Furthermore, vascular endothelial growth factor (VEGF-121), which is downregulated after injury, minimized vascular dropout, but had no effect on the proliferative potential of endothelial cells. These data imply VEGF-121 functions to limit apoptosis and prevent endothelial to mesenchymal transformation, and the resulting fibrosis and microvascular dropout. The important role for endothelial cells after injury in fibrosis is in agreement with data from three

25 Horbelt M, Lee SY, Mang HE, Knipe NL, Sado Y, Kribben A, Sutton TA: Acute and chronic microvascular alterations in a mouse model of ischemic acute kidney injury. Am J Physiol Renal Physiol 2007;293:F688–F695.
26 Basile DP, Friedrich JL, Spahic J, Knipe NL, Mang HE, Leonard EC, Ashtiyani SC, Bacallao RL, Molitoris BA, Sutton TA: Impaired endothelial proliferation and mesenchymal transition contribute to vascular rarefaction following acute kidney injury. Am J Physiol Renal Physiol 2011;300:F721–F733.
27 Zeisberg EM, Potenta SE, Sugimoto H, Zeisberg M, Kalluri R: Fibroblasts in kidney fibrosis emerge via endothelial-to-mesenchymal transition. J Am Soc Nephrol 2008;19:2282–2287.
28 Leonard EC, Friedrich JL, Basile DP: VEGF-121 preserves renal microvessel structure and ameliorates secondary renal disease following acute kidney injury. Am J Physiol Renal Physiol 2008;295:F1648–F1657.
29 Okusa MD, Chertow GM, Portilla D: The nexus of acute kidney injury, chronic kidney disease, and World Kidney Day 2009. Clin J Am Soc Nephrol 2009;4:520–522.
30 Molitoris BA: Contrast nephropathy: are short-term outcome measures adequate for quantification of long-term renal risk? Nat Clin Pract Nephrol 2008;4:594–595.
31 Venkatachalam MA, Griffin KA, Lan R, Geng H, Saikumar P, Bidani AK: Acute kidney injury: a springboard for progression in chronic kidney disease. Am J Physiol Renal Physiol 2010, E-pub ahead of print.
32 Tongers J, Losordo DW: Frontiers in nephrology: the evolving therapeutic applications of endothelial progenitor cells. J Am Soc Nephrol 2007;18:2843–2852.
33 Becherucci, F, Mazzinghi B, Ronconi E, Peired A, Lazzeri E, Sagrinati C, Romagnani P, Lasagni L: The role of endothelial progenitor cells in acute kidney injury. Blood Purif 2009;27:261–270.
34 Akcay A, Nguyen Q, Edelstein CL: Mediators of inflammation in acute kidney injury. Mediators Inflamm 2009;2009:137072.
35 Gluba A, Banach M, Hannam S, Mikhailidis DP, Sakowicz A, Rysz J: The role of Toll-like receptors in renal diseases. Nat Rev Nephrol 2010;6:224–235.
36 Wu H, Chen G, Wyburn KR, Yin J, Bertolino P, Eris JM, Alexander SI, Sharland AF, Chadban SJ: TLR4 activation mediates kidney ischemia/reperfusion injury. J Clin Invest 2007;117:2847–2859.
37 El-Achkar TM, Plotkin Z, Marcic B, Dagher PC: Sepsis induces an increase in thick ascending limb Cox-2 that is TLR4 dependent. Am J Physiol Renal Physiol 2007;293:F1187–F1196.
38 El-Achkar TM, Wu XR, Rauchman M, McCracken R, Kiefer S, Dagher PC: Tamm-Horsfall protein protects the kidney from ischemic injury by decreasing inflammation and altering TLR4 expression. Am J Physiol Renal Physiol 2008;295:F534–F544.
39 Rusai K, Sollinger D, Baumann M, Wagner B, Strobl M, Schmaderer C, Roos M, Kirschning C, Heemann U, Lutz J: Toll-like receptors 2 and 4 in renal ischemia/reperfusion injury. Pediatr Nephrol 2010;25:853–860.
40 Casellas D, Mimran A: Aglomerular pathways in intrarenal microvasculature of aged rats. Am J Anat 1979;156:293–299.
41 Bialestock D: The extra-glomerular arterial circulation of the renal tubules. Anat Rec 1957;129:53–63.

Prof. Bruce A. Molitoris, MD
Division of Nephrology, Department of Medicine
Indiana University School of Medicine
950 W. Walnut Street, R2–202, Indianapolis, IN 46202 (USA)
Tel. +1 317 274 5287, E-Mail Bmolitor@iupui.edu

Kellum JA, Ronco C, Vincent J-L (eds): Controversies in Acute Kidney Injury.
Contrib Nephrol. Basel, Karger, 2011, vol 174, pp 119–128

The Pathogenesis of Acute Kidney Injury and the Toxic Triangle of Oxygen, Reactive Oxygen Species and Nitric Oxide

Ugur Aksu[a,c] · Cihan Demirci[c] · Can Ince[a,b]

[a]Department of Translational Physiology, Academic Medical Center, University of Amsterdam, Amsterdam, [b]Department of Intensive Care, Erasmus MC University Hospital Rotterdam, Rotterdam, The Netherlands; [c]Department of Biology, Faculty of Science, Istanbul University, Istanbul, Turkey

Abstract

Despite the identification of several of the cellular mechanisms thought to underlie the development of acute kidney injury (AKI), the pathophysiology of AKI is still poorly understood. It is clear, however, that instead of a single mechanism being responsible for its etiology, AKI is associated with an entire orchestra of failing cellular mechanisms. Renal microcirculation is the physiological compartment where these mechanisms come together and exert their integrated deleterious action. Therefore, the study of renal microcirculation and the identification of the determinants of its function in models of AKI can be expected to provide insight into the pathogenesis and resolution of AKI. A major determinant of adequate organ function is the adequate oxygen (O_2) supply at the microcirculatory level and utilization at mitochondrial levels for ATP production needed for performing organ function. The highly complex architecture of the renal microvasculature, the need to meet a high energy demand and the borderline hypoxemic nature of the kidney makes it an organ that is highly vulnerable to injury. Under normal, steady-state conditions, the oxygen supply to the renal tissues is well regulated and utilized not only for mitochondrial production of ATP (mainly for Na reabsorption), but also for the production of nitric oxide and the reactive oxygen species needed for physiological control of renal function. Under pathological conditions, such as inflammation, shock or sepsis, however, the renal microcirculation becomes compromised, which results in a disruption of the homeostasis of nitric oxide, reactive oxygen species, and oxygen supply and utilization. This imbalance results in these compounds exerting pathogenic effects, such as hypoxemia and oxidative stress, resulting in further deterioration of renal microcirculatory function. Our hypothesis is that this sequence of events underlies the development of AKI and that integrated therapeutic modalities targeting these pathogenic mechanisms will be effective therapeutic strategies in the clinical environment.

Despite the advances made in the pathogenesis and treatment of acute kidney injury (AKI), current therapeutic modalities have been ineffective for adequate treatment. Consequently, AKI remains a condition with a poor prognosis in hospitals and intensive care units today. Major contributors to the number of patients with AKI are renal ischemia and hypoxia that can occur in clinical situations, such as kidney transplantation, treatment of suprarenal aneurysms, cardiac surgery and renal artery reconstructions, contrast agent-induced nephropathy, cardiac arrest, sepsis, and shock [1]. AKI is associated with higher early and late mortality rates [1] and with higher costs, especially related to the demand for retransplantation and hemodialysis dependency. Among critical care patients who have AKI and survive, up to 30% will require long-term dialysis [2].

AKI, classified according to the RIFLE criteria, is characterized by the sudden loss of glomerular filtration rate (GFR) with an accumulation of nitrogenous waste products. Current opinion concerning the pathophysiology of AKI implicates the reduction in renal blood flow and the consequent renal ischemia as the cause of the loss of GFR that causes disturbances in fluid, electrolyte and acid-base balances [3]. Inflammatory processes can be triggered by ischemic insults, leading to increased expression of adhesion molecules and impaired tubular sodium reabsorption due to intraluminal debris from tubular cells. Endothelial injury directly affects afferent arterioles and results in endothelin release and further vasoconstriction [4], which together cause renal microcirculatory dysfunction.

Measures of biomarkers in the blood and/or urine have recently been developed for diagnosis of AKI at an early stage of damage, and they can potentially be used to prevent progression. Early diagnosis (before GFR falls) is directly related to progression of injury. Increases in serum creatinine and/or blood urea nitrogen are commonly seen in AKI. However, traditional blood and urine biomarkers (such as the fractional excretion of sodium) are nonspecific and not sensitive. New biomarkers have been discovered by using advanced molecular techniques. These biomarkers have been assessed primarily after a specific insult, such as cardiac surgery, kidney transplantation, contrast administration and sepsis. Currently, some urine biomarkers, such as serum neutrophil gelatinase lipocalin (sNGAL), cystatin C, urinary kidney injury molecule, interleukin-18 and NGAL (uNGAL), have been tested for early diagnosis of AKI [5, 6]. Despite these advances in the diagnosis of AKI, however, the pathophysiology and progress of AKI to renal failure remain poorly understood [7].

Oxygen, Reactive Oxygen Species and Nitric Oxide

The kidney utilizes oxygen predominantly for sodium transport, and tissue oxygen tensions are around 45–50 mm Hg in the cortex and between 10–18 mm Hg in the medulla [8]. There is a positive correlation between renal oxygen consumption (VO_2) and tubular sodium reabsorption and GFR [9]. The kidney is highly sensitive to hypoxia because of its microvascular structure [10]. Besides its utilization for mitochondrial respiration, oxygen is also used for reactive oxygen species (ROS) and nitric oxide (NO) generation.

In addition to its direct vasodilatory effect, NO produced by endothelial nitric oxide synthase (NOS) is thought to prevent vascular dysfunction by inhibiting platelet aggregation and preventing leukocyte activation and infiltration via endogenous anti-inflammatory properties. The depletion of suitable cofactors of the NO-producing enzyme (endothelial NOS; e.g. L-arginine and BH4), such as occurs during reperfusion injury and sepsis, can enhance production of ROS by uncoupling endothelial NOS [11]. Excessive NO produced by cells (occurring, for example, as a result of inflammation from inducible NOS activation) can inhibit mitochondrial respiration by competing with oxygen mitochondrial cytochrome oxidase in a dose-dependent manner [12]. Thus, the production and utilization of oxygen, NO and ROS are intrinsically dependent on each other for ensuring adequate renal function.

This delicate homeostasis is pathogenically altered during inflammation and hypoxemia, and leads to oxidative stress and tissue damage. Oxidative stress is an imbalance between oxidants and antioxidants that favors oxidants and causes a disruption of redox signaling and control, leading to damage of cellular molecular structures [13]. Oxygen radicals can be released after the reduction of oxygen, and the outcome is cell injury and dysfunction. ROS is a common term that is used for both oxygen radicals (O_2^- and OH^-) and nonradical (H_2O_2, HOCl, O_3) compounds. Another commonly used term is 'oxidant'. However, O_2^- and H_2O_2 can function as both oxidizing and reducing agents.

Under normal circumstances, ROS are released at low concentrations and are neutralized by endogenous antioxidant compounds, which can be both enzymatic, such as superoxide dismutase, glutathione peroxidase, catalase and glutathione peroxidase, and nonenzymatic, such as glutathione and vitamins C and E. Both high and low levels of oxygen tension promote oxidative stress, making the need for keeping levels of tissue oxygen tensions at physiological levels imperative for avoiding the detrimental effects of oxidative stress [13].

Several studies have promoted the idea that targeting the ROS associated with cellular injury in acute or chronic kidney disease may aid the design of future therapeutic approaches. Oxidative stress commonly results in the degeneration of cells via apoptotic pathways in the kidney. Apoptotic-induced oxidative stress in conjunction with processes of mitochondrial dysfunction form the corner stone of triggered mechanisms in nephropathic conditions. The dependency of ROS activity on oxygen availability was recently shown in a model of oxidative stress in spontaneously hypertensive rats where a loss of bioactive NO by high ROS production was found to interfere with normal oxygen usage in the kidney. In addition, it was shown that superoxide produced by NADPH oxidase is inhibited when oxygen tensions drop below 20 mm Hg [14].

The main sources of ROS in the microcirculation are mitochondria, NADPH oxidase, NO synthase and xanthine oxidase. Moreover, cytochrome P450 and cyclooxygenase are capable of producing O_2^-. In mitochondria, there is continued production of ROS during cellular respiration. A percentage of the oxygen used in mitochondria is reduced to superoxide. This process occurs by blockade of the electron transfer chain at the flavin mononucleotide group of complex I or at the ubiquinone site of complex III. This free radical generation is under the control of the endogenous antioxidant defense system, and Mn-superoxide dismutase in mitochondria converts the superoxide to H_2O_2. Superoxide generates much of its biological effects by scavenging the NO produced by three isoforms of NOS, each expressed in the kidney: neuronal NOS, inducible NOS and endothelial NOS.

In their pathogenic action, ROS mostly cause their deleterious effect by inducing lipid peroxidation, activation of apoptotic pathways, alteration of intracellular calcium concentrations and inducement of adhesion molecule expression. Oxidative stress can also cause mitochondrial membrane permeability, resulting in loss of mitochondrial NAD+ residues and subsequent radical generation [15]. In addition to their pathogenic action, superoxide and NO are involved in normal kidney and vascular functions. Both may mediate the maintenance of vascular tone (especially in the afferent arterioles) and tubular function. Angiotensin II mediated by superoxide together with NO, is also responsible for maintaining afferent arteriole tone in perfused isolated mouse afferent arterioles. Renal oxygen consumption, by contrast, has been found to be increased by L-NG-monomethyl-arginine, a nonselective NOS inhibitor, and S-methyl-L-thiocitrulline, a selective neuronal NOS inhibitor [16], which implicates the different isoforms of NOS in modulating oxygen utilization. These and the above-mentioned studies illustrate the complicated interdependency between oxygen and NO; their homeostasis becomes severely altered during conditions of renal inflammation and hypoxemia-reperfusion, resulting in oxidative stress and loss of renal function.

Sepsis, Fluid Therapy and Acute Kidney Injury

Currently, sepsis is one of the main contributors to the development of AKI in intensive care [3]. Sepsis constitutes a major hit to renal function in that it induces a massive increase in inflammatory mediators and activated leukocytes, which together cause severe microcirculatory dysfunction and disruption of oxygen homeostasis that leads to oxidative stress and hypoxemia [17]. In the early stages of sepsis, impairment of renal microcirculation is a key complication potentially leading to AKI through hypoxia-induced tubular epithelial cell injury and acute tubular necrosis.

Fluid resuscitation is a cornerstone of the treatment of sepsis because it is considered crucial for the preservation of adequate intravascular volume and the maintenance of blood pressure [18]. Such fluid therapy is expected to promote microvascular perfusion and thereby renal oxygenation. However, the extent to which fluid therapy is effective in promoting renal oxygenation has recently been questioned [19]. Fluid resuscitation can have severe deleterious effects on the microcirculation [20] and hemodilution in a range of therapeutic scenarios have been found to lead to renal failure [21]. Excessive fluid administration in sepsis has been found to be associated with renal failure [22], although restrictions in fluid use can lead to hypovolemia. Therefore, determining the optimal fluid volume to administer during sepsis to deal with hypovolemia remains a source of controversy [20]. In addition, the type of fluid that yields the best renal outcome when used for resuscitation in sepsis is also currently a source of uncertainty. This controversy not only includes the use of crystalloid versus colloid solutions, but also encompasses the use of balanced versus unbalanced colloid solutions. Because most colloid preparations are saline-based, liberal fluid resuscitation regimes may lead to non-physiologically high sodium and chloride concentrations and may be associated with the development of (hyperchloremic) metabolic acidosis that can affect inflammatory and coagulation homeostasis, thereby contributing to the deterioration of renal function. This insight has led to development of modern hydroxyl-ethylated starch (HES) preparations, based on balanced plasma-adapted crystalloid solutions, and to the notion of developing a totally balanced fluid resuscitation concept, including balanced crystalloids and balanced colloids. Studies have also found that HES solutions have anti-inflammatory properties [23]. In contrast, however, various investigations have identified potential adverse effects of HES solutions on renal function.

In a study we investigated the effects on renal microcirculation of a total balanced volume replacement regimen, including a Ringer's acetate-balanced HES preparation, in comparison to an unbalanced fluid regimen. To this end, we employed a rat model to study alterations in renal cortex microcirculatory perfusion associated with LPS-induced endotoxemia [24]. We monitored kidney cortical perfusion using laser speckle imaging [25]. This system is noninvasive and macroscopic (but sensitive to microcirculatory flow), quantitative and allows scoring of microcirculatory perfusion heterogeneity [25]. In the study, we showed that of the different solutions used for fluid resuscitation, the balanced HES solution was the best for promoting renal cortex perfusion, illustrating

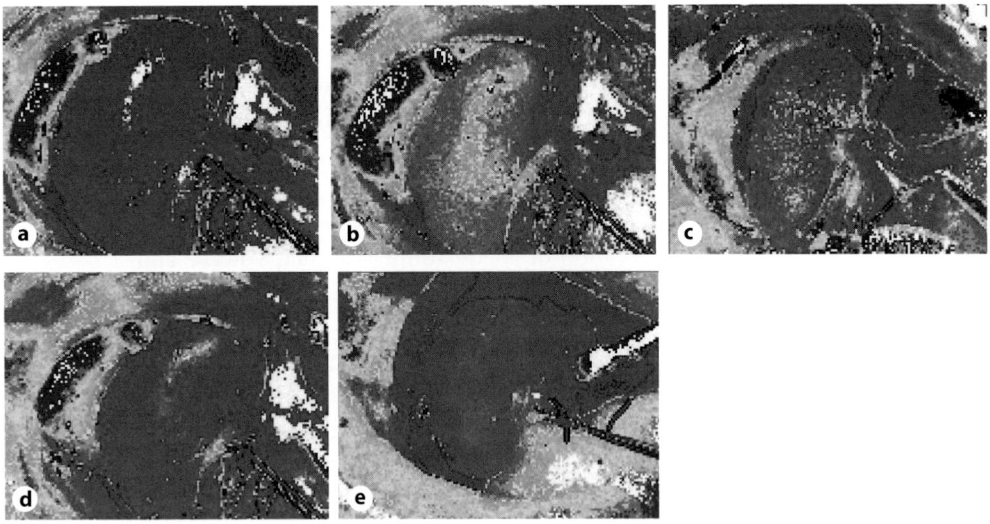

Fig. 1. Perfusion of the renal cortex shown by speckle imaging following fluid resuscitation during endotoxemia. Speckle images at control (**a**), during endotoxemia (**b**) and after resuscitation with 0.9% NaCl (**c**), HES-0.9%NaCl (**d**) and HES Ringer's acetate (**e**). Red denotes good perfusion, whereas green and blue denote progressively less perfusion. As can be seen, NaCl resuscitation causes markedly poor perfusion in comparison to a balanced starch solution, such as HES Ringer's acetate.

how the choice of fluids for resuscitation in sepsis may effect renal perfusion (fig. 1).

In our model, hyperchloremia and the resulting metabolic acidosis were strongly associated with increased perfusion heterogeneity in these endotoxemic animals. Although the mechanisms responsible for the observed increase in renal vascular resistance due to hyperchloremic metabolic acidosis are relatively well understood, the mechanisms leading to increased perfusion heterogeneity are not. Acidemia may affect a variety of vasoregulatory mechanisms. For example, acidemia may influence the inflammatory response through at least two mechanisms: by increasing endogenous catecholamine release, which induces the release of both pro- and anti-inflammatory cytokines; and/or by NO having a direct effect on the renal microvasculature. Increased levels of tumor necrosis factor may also influence perfusion heterogeneity, either by the vasoactive properties of these molecules or via direct tissue injury.

Despite much literature showing its deleterious effects, 0.9% NaCl is widely used as a resuscitation solution in emergency departments and intensive care units today. However, it is known that crystalloids have a poor plasma expander effect since they rapidly leave the intravascular space. Compared to colloids in this respect, additional (about three times more) volumes of crystalloids need to be given to reach a similar systemic hemodynamic endpoint [26]. Therefore, colloid fluids are more effective as plasma expanders because less is needed, although colloids have been shown to have adverse affects on coagulation pathways and are often dissolved in high chloride solutions (e.g. 0.9% NaCl). Excessive Cl^- levels can have a deleterious effect on renal function. For example, intrarenal administration of a chloride solution provokes renal vasoconstriction and reduces GFR [27]. Solutions containing high amounts of Cl^- can cause hyperchloremic acidosis, while solutions with buffers, such as acetate, and more physiological concentrations of strong ions do not. Related mechanisms involving acidosis have been clearly described [28].

Conclusions

It is clear that physiological function of the kidney relies on a delicate balance between oxygen transport and utilization, reactive oxygen and NO metabolism, and that this balance results in effective renal microcirculation that is essential for renal function. Inflammation and hypoxemia can severely disrupt this delicate balance, resulting in oxidative stress and regional hypoxemia that lead to renal failure. Sepsis and reperfusion injury are the most severe manifestations of such an inflammatory insult attacking microcirculatory function at all levels; they can result in a viscous, self-perpetuating spiral of pathogenic events that lead to renal failure. In this sequence of events, activated leukocytes and inflammatory mediators disrupt the homeostasis of renal oxygenation, which leads to functional deterioration. This model of the pathogenesis of AKI can be seen in figure 2. According to this hypothesis of the progress of AKI, therapeutic compounds need to correct all the elements of this pathology (e.g. inflammation, microcirculatory oxygenation, NO and oxidative stress) in an integrated manner to effectively alter the course of AKI. We performed empirical studies on compounds that have multiple correcting effects on these pathogenic mechanisms (e.g. dexamethasone, L-NIL, iloprost and APC;), and we found them successful in reverting AKI in rat models of AKI [29–31]. The question now is whether the approach based on the integrative model provides a strategic therapeutic rationale for the treatment of AKI in the clinical scenarios.

27 Wilcox CS: Regulation of renal blood flow by plasma chloride. J Clin Invest 1983;71:726–735.
28 Kellum JA, Song M, Li J: Science review: extracellular acidosis and the immune response: clinical and physiologic implications. Crit Care 2004;8:331–336.
29 Johannes T, Mik EG, Klingel K, Dieterich HJ, Unertl KE, Ince C: Endotoxin-induced acute renal failure reversed by low-dose dexamethasone. Shock 2009;31:521–528.
30 Legrand M, Almac E, Mik EG, Johannes T, Kandil A, Bezemer R, Payen D, Ince C: L-NIL prevents renal microvascular hypoxia and increase of renal oxygen consumption after ischemia-reperfusion in rats. Am J Physiol Renal Physiol 2009;296:F1109–F1117.
31 Johannes T, Ince C, Klingel K, Unertl KE, Mik EG: Iloprost restores renal oxygenation and kidney function in endotoxemia-related acute renal failure in the rat. Crit Care Med 2009;37:1423–1432.

Dr. Can Ince
Department of Translational Physiology
Academic Medical Center, University of Amsterdam
Amsterdam (The Netherlands)
E-Mail c.ince@amc.uva.nl

Kellum JA, Ronco C, Vincent J-L (eds): Controversies in Acute Kidney Injury.
Contrib Nephrol. Basel, Karger, 2011, vol 174, pp 129–137

Pathogenesis of Acute Kidney Injury: Effects of Remote Tissue Damage on the Kidney

Xiaoyan Wen · Zhiyong Peng · John A. Kellum

The Clinical Research, Investigation, and Systems Modeling of Acute illness (CRISMA) Laboratory, Department of Critical Care Medicine, University of Pittsburgh, Pittsburgh, Pa., USA

Abstract

The modern definition and classification of acute kidney injury (AKI) has now been applied to thousands of patients around the world and in different settings. Epidemiology is shedding intense light on the credibility of our fundamental notions of how AKI occurs and why. It is clear from multiple studies that sepsis is the leading etiology of AKI, although other settings associated with systemic inflammation (polytrauma, burns, pancreatitis, cardiopulmonary bypass) also represent important means of exposure. Early in sepsis-induced AKI, there is an intense reduction in glomerular filtration with only mild structural changes mostly limited to the tubular epithelium. Although endothelial and interstitial changes may also occur, histopathology in sepsis-induced AKI is characteristically bland. Of course, many patients with severe AKI do not recover (as many as 50% in recent studies), and these patients ultimately develop severe structural alterations including fibrosis. Recent evidence suggests that cells in injured and infected tissues release immunological danger signals or danger-associated molecular patterns which communicate with remote organs including the kidney, where they activate dendritic cells and T cells and thus initiate inflammation. Dissonance of mediator secretion and cell responses may lead to persistent injury and de novo chronic kidney disease. A number of soluble mediators initiate a variety of pathophysiological processes as kidney injury evolves. In this chapter, we will discuss the pathogenesis of AKI in light of new information concerning injury and repair, and focus on the controversies arising from emerging evidence.

Acute kidney injury (AKI) occurs in up to 70% of critically ill patients and mortality is more than twice that seen in similar patients without AKI [1]. A multicenter intensive care unit study found that sepsis accounted for 32.4%

of all patients with AKI, and was associated with greater severity of AKI and higher mortality [2]. Hence, understanding the pathogenesis of sepsis-induced AKI, including developing tools for early diagnosis, is pivotal to improve outcomes.

Despite the strong association between sepsis and AKI, the precise mechanisms responsible for functional deterioration are still unclear and therapies are mainly supportive. However, recent studies suggest that leukocyte activation and upregulation of soluble mediators such as cytokines and chemokines, as well as their interactions, are key factors in the pathogenesis of AKI, its progression and the development of complications. In a prospective multicenter study of 1,836 patients with community-acquired pneumonia, Murugan et al. [3] reported that 34% of patients with pneumonia developed AKI. The occurrence, severity and outcomes of AKI were significantly associated with high concentrations of plasma cytokines, especially interleukin 6 (IL-6) [3]. Moreover, the majority of patients with sepsis-induced AKI in this study were not admitted to an intensive care unit and were not in shock. Moreover, animal models of AKI that rely on ischemia-reperfusion bare little resemblance to sepsis-induced AKI in humans and produce histopathologic changes not seen in humans [4]. Accordingly, the generation of new information from relevant models and clinical material leading to new paradigms are urgently needed.

In this review we will argue that (1) injury to the kidney is largely derived from local inflammation, (2) the triggers for this inflammation are derived from remote tissue damage, and (3) treatment directed at blocking damage signals or suppressing the response to them can attenuate kidney injury.

The Kidney and Inflammation

Sepsis-induced AKI, while being the most common cause of AKI in humans, may be a special form of the disease, typified by a systemic inflammatory state in which the 'primary disease' is usually not in the kidney. Although renal biopsies are seldom performed in cases of septic AKI, available data from autopsies and animal studies suggest that sepsis-induced AKI is largely an acellular process with little evidence of tubular necrosis [5]. However, a strong association with proinflammatory cytokine expression and the development of AKI [3, 6] suggests an important role of systemic inflammatory mediators in the process of septic AKI.

Like all epithelia, renal tubular cells are an actively replenishing population and have the capacity to respond to injury by undergoing dedifferentiation, migration, proliferation and redifferentiation. Tubular epithelial cells acquire the ability to migrate and proliferate only after they have dedifferentiated into a progenitor phenotype. Importantly, the timing and direction of cell transition is largely determined by the counterbalance of several growth factors, adhesion molecules and cytokines including transforming growth factor beta-1 (TGF-β1) and hepatocyte growth factor (for more details see Wen et al. [7]). Genetic 'fate-mapping' techniques show that the majority of regenerated tubule epithelial cells come from injured intrinsic tubular cells [8]. Proximal tubular cells are also able to respond to various stimuli by producing a variety of inflammatory mediators [9–13]. Recent evidence suggests that IL-6 enhances TGF-β1 signaling by reinforcing sustained endocytosis [14], which may help explain the association between plasma IL-6 levels and the adverse outcome we observe in AKI in humans [3].

Resident dendritic cells and T cells respond to a variety of locally and remotely produced molecular signals. Cytokines/chemokines promote intense inflammation in local injured regions through activated Toll-like receptors (TLRs), , complement cascade and nuclear factor kappa-B (NF-κB). The initial injury and proinflammatory state may lead to mononuclear cell infiltration and macrophage activation, and promote angiogenesis and fibrous tissue deposition.

Another source of inflammation was suggested by Paladino et al. [15] who noted that leukocytes passing through the kidney are exposed to a uniquely hostile environment since the renal medulla is severely hypertonic (approx. 800 mOsm/kg) and hypoxic (tissue PO2 approx. 20 Torr). These authors suggest that leukocytes transiting this hostile environment might well be 'primed' to respond to any second signal provided by the renal endothelium; such signals might promote activation of leukocytes or changes in leukocyte adhesion molecule expression. Thus, local propagation of the inflammatory response is plausible within the kidney due to the complex interplay of dendritic cells, renal tubular epithelium, and transiting leukocytes and their interaction with the endothelium (fig. 1).

Finally, although still controversial [16–18], there is evidence that the various actors in this drama can change roles. Experimental data show that intrinsic renal cells can undergo phenotypic transformation to progenitor cells under the influence of the vary mediators such as inflammatory cytokines and growth factors that are generated from the interactions described above. Dedifferentiation from renal tubular epithelial cells to mesenchymatous cells is the best documented cell transition in the kidney, and has been termed 'epithelial-mesangial transdifferentiation' [16]. Dedifferentiated epithelial cells lose cell polarity and cell-cell junctions, detach from the basement membrane, and lose their normal phenotype and function. They acquire new characters of both muscle cells and fibroblasts called myofibroblasts. These activated fibroblasts are capable of proliferation, migration and synthesis of extracellular matrix, and can play integral roles in both repair and fibrosis. Furthermore, studies have shown that tubular epithelial cells can transform into other cell types such as phagocytes, which are also characterized by the expression of proteins such as kidney injury molecule-1 (KIM-1) [19]. These cells acquire the ability to internalize adjacent apoptotic bodies and necrotic cells, and remove irreversibly damaged tubular

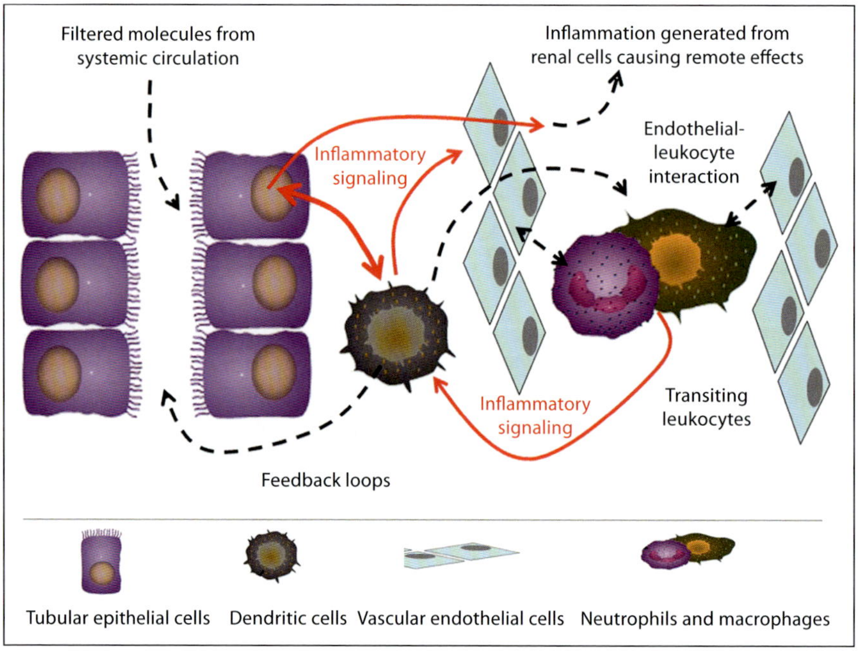

Fig. 1. Filtered damage molecules and inflammatory mediators stimulate epithelial and endothelial cells which in turn signal dendritic cells with activation of other immune effector cells. Transiting neutrophils and monocytes interact with the endothelium and further propagate inflammation. Dashed black arrows denote incoming signals and red solid arrows show inflammatory mediator release.

cells in absences of infiltrated macrophages [19]. Thus, in sepsis, injury to the kidney is largely derived from local inflammation.

Ultimately, unresolved inflammation can lead to severe structural changes as the delicate balance between deposition of extracellular matrix components and enzymatic degradation is altered [20]. Both injured mesangial cells and endothelial cells eventually undergo apoptosis. Disturbances involving anatomic/functional changes or concomitant diseases interfering with soluble mediator secretion, cell-cell communication, and cellular phenotype transition will lead to ongoing inflammation and extensive apoptosis. Intrinsic cell loss, glomerulosclerosis and progressive interstitial fibrosis may progress to result in incomplete renal recovery or even organ death.

Remote Triggers of Kidney Inflammation

An adult human kidney produces some 100–200 l of ultrafiltrate daily, reabsorbing more than 98%. The proximal tubular epithelial cells are exposed to numerous substances including small peptides, hormones, inflammatory molecules and nitrogenous waste products. These cells also actively secrete molecules from the peritubular capillary bed into the tubular lumen providing an alternate route of clearance for substance that are not filtered at the glomerulus. As such, the proximal tubular epithelial cells are uniquely positioned to respond to toxins and mediators coming from remote tissues or exogenously.

A classic example of remote pathology causing AKI is rhabdomyolysis. Myoglobin causes AKI through a combination of oxidative injury, vascular effects leading to medullary hypoperfusion and hypoxia, and tubular obstruction [21]. However, local inflammation also magnifies this injury. In addition, a number of studies over many years (reviewed in [22]) have implicated other factors in the pathogenesis of myonecrosis-induced AKI, factors which may lead to renal injury through multiple different mechanisms. At the very least, rhabdomyolysis represents a clear example of remote tissue damage leading to AKI through a variety of different pathways. Another example of 'heme-induced' AKI may be what occurs in the setting of cardiopulmonary bypass [23]. Interestingly, hemolysis itself is an uncommon cause of AKI, suggesting that factors other than heme exposure per se are important in the pathogenesis of AKI in such settings.

Injury to the lung has been implicated as a cause of AKI. Animal models have shown alterations in distant organs in response to lung injury and mechanical ventilation. For example, large tidal volume ventilation, alone or in conjunction with acid aspiration, has been shown to alter the concentrations of numerous cytokines [24, 25] and chemokines [26] in the kidney, as well as affect inflammatory cell distribution and apoptosis [26]. These results would seem to substantiate the principle that remote organ injury leads to AKI through modulation of intrarenal inflammation. However, the picture is less clear as a subsequent study by Hoag et al. [27] found no effects on the kidney in dogs during acute lung injury when gas exchange and hemodynamics were carefully monitored. In its totality, this literature seems to suggest that multiple factors contribute to the development of AKI in experimental models, a scenario reminiscent of clinical AKI. However, perhaps the best example of remote processes causing AKI comes from human sepsis. In a recent in vitro study, plasma from septic patients was shown to induce granulocyte adhesion, apoptosis and altered polarity in tubular cells in culture [28].

Regardless of the type of tissue injury, injured cells release immunological danger signals or danger-associated molecular pattern molecules (DAMPs) such as the chromatin-associated protein high-mobility group box 1 (HMGB1), DNA, micro RNAs, ATP, adenosine and uric acid. These molecules activate proximal tubule cells which respond by producing cytokines [9–13]. Immune effector cells such as dendritic cells also directly respond to DAMPs, but also indirectly respond to mediators produced by the epithelium. Dendritic cells are constantly in communication with other cells capable of priming and activating the immune system for response against various insults. Several important inflammatory molecules and pathways can become involved, including TLRs and NF-κB. The complement cascade may also become activated as part of DAMP signaling.

Finally, it is important to point out that epithelial cells are not alone in their ability to respond to signals from remote sources. Chen et al. [29] recently reported results implicating endothelial Toll-like receptor 4 (TLR4) in the initiation of endothelial cell activation in ischemic AKI. These findings further support the idea that the innate immune system plays an important role in the pathogenesis of ischemic AKI [30]. Moreover, if signals from remote injury and infection can lead to both vascular and epithelial changes, a unifying paradigm of AKI in sepsis may be emerging.

Taken together, the picture is one of multiple costimulatory effects and feedback loops within the kidney and extending beyond (fig. 1). Filtered DAMPs and inflammatory mediators stimulate epithelial and endothelial cells which in turn signal dendritic cells with activation of T cells and other cells. Transiting neutrophils and monocytes need not infiltrate into the renal parenchyma to exert effects. The culmination of these multiple signals is organ damage. If the signals resolve quickly, the process can be reversed, but if they remain and intensify, phenotypic transformation will lead to fibrosis and permanent loss of function.

Novel Therapies

We propose that treatment directed at blocking damage signals or suppressing the response to them can attenuate kidney injury. Cantaluppi et al. [28] used a sorbent (Amberchrom CG161 M resin) to remove mediators from the plasma of septic humans in vitro and compared it to untreated septic plasma in terms of the effects on renal tubular epithelial cells in culture. Plasma adsorption significantly reduced the concentrations of several soluble mediators and decreased renal cell injury. The inhibition of granulocyte adhesion to tubular cells was associated with the downregulation of intercellular adhesion molecule 1 (ICAM-1) and CD40. Adsorption inhibited tubular cell apoptosis induced by septic plasma by downregulating the activation of caspase-3, -8 and -9 as well as Fas/death receptor-mediated signaling pathways. Numerous effects induced by cell exposure to septic plasma were all significantly reduced by resin adsorption, including alteration of cell polarity, morphogenesis and protein reabsorption, as well as the downregulation of the tight junction molecule ZO-1 (zonula occludens-1), proximal tubular sodium transporter sodium-hydrogen exchanger-3 (NHE3), glucose transporter 2 (GLUT-2) and the endocytic receptor megalin [28]. We have preliminarily shown in vivo using a rodent model of sepsis that removal of various mediators using a hemoadsorption system (CytoSorb) results in protection against sepsis-induced

AKI and liver damage [31]. Furthermore, there was a significant reduction in the plasma concentration of HMGB1, a potent and well-known DAMP.

A host of other potential treatments may be on the horizon. Antibody to TGF-β, angiotensin-converting enzyme inhibition, adenosine 2A receptor agonists, and Toll-like receptor inhibition (e.g. chloroquine) [32] may be viable ways to attenuate the inflammatory response in the kidney and ameliorate renal damage. We have preliminarily shown that a single dose of cyclosporine A protects mice from severe AKI induced by folic acid [33]. We found reduced tubular cell injury and interstitial inflammatory cell infiltration. Cyclosporine A protected renal tubular cells from apoptosis, suppressed inflammatory cytokines, abolished upregulation of TNF-related weak inducer of apoptosis (TWEAK) as well as NF-κB activation, and inhibited subsequent macrophage and neutrophil infiltration. As a result of these actions, cyclosporine A decreased interstitial fibrosis and preserved kidney structure and function [33].

Conclusion

Recent evidence from multiple laboratories suggests that cells in injured and infected tissues release immunological danger signals or DAMPs which communicate with remote organs including the kidney. DAMPs activate dendritic cells and T cells and thus initiate inflammation. Emerging evidence supports the idea that injury to the kidney is largely derived from local inflammation triggered by remote tissue damage. Novel treatment strategies are being developed which are directed at blocking damage signals or suppressing the response to them. The hope is that these measures can attenuate kidney injury or shorten its duration, leading to enhanced chances for recovery.

References

1 Ricci Z, Cruz D, Ronco C: The RIFLE criteria and mortality in acute kidney injury: a systematic review. Kidney Int 2008;73:538–546.

2 Bagshaw SM, George C, Dinu I, Bellomo R: A multi-centre evaluation of the RIFLE criteria for early acute kidney injury in critically ill patients. Nephrol Dial Transplant 2008;23:1203–1210.

3 Murugan R, Karajala-Subramanyam V, Lee M, Yende S, Kong L, Carter M, Angus DC, Kellum JA: Acute kidney injury in non-severe pneumonia is associated with an increased immune response and lower survival. Kidney Int 2010;77:527–535.

4 Heyman SN, Rosenberger C, Rosen S: Experimental ischemia-reperfusion: biases and myths-the proximal vs. distal hypoxic tubular injury debate revisited. Kidney Int 2010;77:9–16.

5 Langenberg C, Bagshaw SM, May CN, Bellomo R: The histopathology of septic acute kidney injury: a systematic review. Crit Care 2008;12:R38.

6 Bagshaw SM, Langenberg C, Haase M, Wan L, May CN, Bellomo R: Urinary biomarkers in septic acute kidney injury. Intensive Care Med 2007;33:1285–1296.

7 Wen X, Murugan R, Peng Z, Kellum JA: Pathophysiology of acute kidney injury: a new perspective; in Ronco C, Bellomo R, McCullough PA (eds): Cardiorenal Syndromes in Critical Care. Contrib Nephrol. Basel, Karger, 2010, vol 165, pp 39–45.

8 Humphreys BD, Valerius MT, Kobayashi A, Mugford JW, Soeung S, Duffield JS, McMahon AP, Bonventre JV: Intrinsic epithelial cells repair the kidney after injury. Cell Stem Cell 2008;2:284–291.

9 Waeckerle-Men Y, Starke A, Wahl PR, Wüthrich RP: Limited costimulatory molecule expression on renal tubular epithelial cells impairs T cell activation. Kidney Blood Press Res 2007;30:421–429.

10 Jevnikar AM, Brennan DC, Singer GG, Heng JE, Maslinski W, Wuthrich RP, Glimcher LH, Kelley VE: Stimulated kidney tubular epithelial cells express membrane associated and secreted TNF alpha. Kidney Int 1991;40:203–211.

11 Schmouder RL, Strieter RM, Wiggins RC, Chensue SW, Kunkel SL: In vitro and in vivo interleukin-8 production in human renal cortical epithelia. Kidney Int 1992;41:191–198.

12 Nechemia-Arbely Y, Barkan D, Pizov G, Shriki A, Rose-John S, Galun E, Axelrod JH: IL-6/IL-6R axis plays a critical role in acute kidney injury. J Am Soc Nephrol 2008;19:1106–1115.

13 Wu MC, Reuben PM: Granulocyte-macrophage colony-stimulating factor from cultured normal rat kidney cell line. Exp Hematol 1984;12:267–276.

14 Carvajal G, Rodríguez-Vita J, Rodrigues-Díez R, Sánchez-López E, Rupérez M, Cartier C, Esteban V,Ortiz A, Egido J, Mezzano SA, Ruiz-Ortega M: Angiotensin II activates the Smad pathway during epithelial mesenchymal transdifferentiation. Kidney Int 2008;74:585–595.

15 Paladino JD, Hotchkiss JR, Rabb H: Acute kidney injury and lung dysfunction: a paradigm for remote organ effects of kidney disease? Microvasc Res 2009;77:8–12.

16 Kalluri R, Neilson EG: Epithelial-mesenchymal transition and its implications for fibrosis. J Clin Invest 2003;112:1776–1784.

17 Kriz W, Kaissling B, Le Hir M: Epithelial-mesenchymal transition (EMT) in kidney fibrosis: fact or fantasy? J Clin Invest 2011;121:468–474.

18 Duffield JS, Benjamin D Humphreys BD: Origin of new cells in the adult kidney: results from genetic labeling techniques. Kidney Int 2011;79:494–501.

19 Ichimura T, Asseldonk EJ, Humphreys BD, Gunaratnam L, Duffield JS, Bonventre JV: Kidney injury molecule-1 is a phosphatidylserine receptor that confers a phagocytic phenotype on epithelial cells. J Clin Invest 2008;118:1657–1668.

20 Eddy AA: Molecular basis of renal fibrosis. Pediatr Nephrol 2000;15:290–301.

21 Heyman SN, Rosen S, Fuchs S, Epstein FH, Brezis M: Myoglobinuric acute renal failure in the rat: a role for medullary hypoperfusion, hypoxia, and tubular obstruction. J Am Soc Nephrol 1996;7:1066–1074.

22 Poels PJ, Gabreëls FJ: Rhabdomyolysis: a review of the literature. Clin Neurol Neurosurg 199395:175–192.

23 Haase M, Haase-Fielitz A, Bagshaw SM, Ronco C, Bellomo R: Cardiopulmonary bypass-associated acute kidney injury: a pigment nephropathy?; in Ronco C, Bellomo R, Kellum JA (eds): Acute Kidney Injury. Contrib Nephrol. Basel, Karger, 2007, vol 156, pp 340–353.

24 Gurkan OU, O'Donnell C, Brower R, Ruckdeschel E, Becker PM: Differential effects of mechanical ventilatory strategy on lung injury and systemic organ inflammation in mice. Am J Physiol Lung Cell Mol Physiol 2003;285:L710–L718.

25 Rabb H, Wang Z, Nemoto T, Hotchkiss J, Yokota N, Soleimani M: Acute renal failure leads to dysfunction of lung salt and water channels. Kidney Int 2003;63:600–606.

26 Imai Y, Parodo J, Kajikawa O, de Perrot M, Fischer S, Edwards V, Cutz E, Liu M, Keshavjee S, Martin TR, Marshall JC, Ranieri VM, Slutsky AS: Injurious mechanical ventilation and end-organ epithelial cell apoptosis and organ dysfunction in an experimental model of acute respiratory distress syndrome. JAMA 2003;289:2104–2112.

27 Hoag JB, Liu M, Easley RB, Britos-Bray MF, Kesari P, Hassoun H, Haas M, Tuder RM, Rabb H, Simon BA: Effects of acid aspiration-induced acute lung injury on kidney function. Am J Physiol Renal Physiol 2008;294:F900–F908.
28 Cantaluppi V, Weber V, Lauritano C, Figliolini F, Beltramo S, Biancone L, De Cal M, Cruz D, Ronco C, Segoloni GP, Tetta C, Camussi G: Protective effect of resin adsorption on septic plasma-induced tubular injury. Critical Care 2010;14:R4.
29 Chen J, John R, Richardson JA, Shelton JM, Zhou XJ, Wang Y, Wu QQ, Hartono JR, Winterberg PD, Lu CY: Toll-like receptor 4 regulates early endothelial activation during ischemic acute kidney injury. Kidney Int 2011;79:288–299.
30 Jang HR, Rabb H: The innate immune response in ischemic acute kidney injury. Clin Immunol 2009;130:41–50.
31 Peng Z, Wang H, Carter M, DiLeo M, Kellum JA: Hemoadsorption improves long-term survival after sepsis in the rat. Crit Care Med 2008;36:A1.
32 Yasuda H, Leelahavanichkul A, Tsunoda S, Dear JW, Takahashi Y, Ito S, Hu X, Zhou H, Doi K, Childs R, Klinman DM, Yuen PS, Star RA: Chloroquine and inhibition of Toll-like receptor 9 protect from sepsis-induced acute kidney injury. Am J Physiol Renal Physiol 2008;294:F1050–1058.
33 Wen X, Peng Z, Li Y, Wang H, Singbartl K, Kellum JA: Could cyclosporine be renal protective in acute kidney injury? J Am Soc Nephrol 2010;21:342A.

John A. Kellum
Department of Critical Care Medicine
University of Pittsburgh
Pittsburgh, PA 15261
Tel. +1 412 647 7810, E-Mail kellumja@ccm.upmc.edu

Kellum JA, Ronco C, Vincent J-L (eds): Controversies in Acute Kidney Injury.
Contrib Nephrol. Basel, Karger, 2011, vol 174, pp 138–148

A Role for Oxidative Stress

Samuel N. Heyman[a] · Seymour Rosen[b] · Christian Rosenberger[c]

[a]Department of Medicine, Hadassah Hebrew University Hospital, Mt. Scopus, Jerusalem, Israel; [b]Department of Pathology, Beth Israel Deaconess Medical Center and Harvard University, Boston, Mass., USA; [c]Department of Nephrology, Charité Campus Mitte, Berlin, Germany

Abstract

The pathogenesis of sepsis-induced acute kidney injury (AKI) is not fully understood, and may involve altered systemic hemodynamics and renal circulation, renal hypoxia and perhaps direct tubular toxicity. Oxidative stress, induced by systemic and intrarenal generation of reactive oxygen species (ROS) can directly exert renal parenchymal damage and may intensify renal microvascular and functional dysregulation, with a feed-forward loop of hypoxia and ROS generation. Herein we review compelling evidence that sepsis is associated with systemic and intrarenal intense oxidative and nitrosative stress with a depletion of antioxidant capacity. Inflammation and tissue hypoxia conceivably play a central role in the generation of ROS in the septic patient. Though oxidative and nitrosative stress are likely to participate in the pathogenesis of sepsis-induced AKI, it is impossible to clearly identify their isolated independent role and renal-specific effect since there are complex interactions involved linking various affected organs, ROS generation with altered systemic hemodynamics, compromised microcirculation, hypoxia and distorted cellular function. Facing this complex disease entity, alleviation of oxidative stress single-handedly is unlikely to be effective in the prevention of sepsis-associated renal dysfunction. However, the addition of antioxidants to a comprehensive treatment strategy seems a reasonable approach.

Acute kidney injury (AKI) is a complex clinical condition of multiple etiologies, where renal hypoperfusion, altered intrarenal microcirculation and direct tubular toxicity play a role alone or in concert. Since oxidative stress has been implicated in diverse AKI settings, it has been suggested that the inclusion of treatment modalities alleviating oxidative stress could play a role in AKI prevention [1]. Sepsis-associated renal dysfunction is a leading cause of AKI among hospitalized patients, and it is believed that oxidative stress is an especially important factor in this relationship, along with inflammation, altered renal microcirculation and hypoxia [2].

In this review we shall outline the evidence for increased oxidative stress that might participate in the evolution of sepsis-induced AKI. Such oxidative stress may reflect enhanced systemic and intrarenal generation of reactive oxygen species (ROS) in septic patients, as well as hampered antioxidant capacity. We shall also examine the data regarding attempts to attenuate sepsis-induced renal dysfunction with antioxidants, a strategy that if proven effective might underscore the importance of ROS in the pathogenesis of this disorder.

Background

Oxidative/nitrosative stress reflects the net balance of the formation of reactive oxygen and nitrogen species (herein collectively termed oxidative stress and ROS, respectively, for simplicity), and the capacity of the antioxidant systems. ROS is generated through diverse pathways including the mitochondrial electron transport mechanism; the P450 systems; monoamine-, glucose-, xanthine- and NADPH-oxidases; arachidonic acid oxidases; and nitric oxide synthase (NOS) isoforms. Mitochondrial generation forms the major source of renal parenchymal ROS formation, specifically in tubular cells within the hypoxic medulla where some 1–2% of consumed oxygen is normally converted to superoxide. Superoxide-dismutase transforms superoxide to hydrogen peroxide, which is further converted to hydroxyl radical and to peroxynitrite following binding with nitric oxide. ROS are removed by scavengers such as catalase (which converts hydrogen peroxide to water) and are inactivated by a host of antioxidant compounds, mainly proteins and small molecules containing sulfhydryl groups, such as glutathione (GSH). GSH, a ubiquitous major antioxidant, is predominantly synthesized in the liver, which also serves as its major storage site. GSH can be translocated across cell membranes and released into the bloodstream to be taken up and used by other tissues, where it scavenges ROS or restores sulfhydryl groups in other antioxidant molecules. In this process, catalyzed by selenium, oxidized GSH (GSSG) is formed. Its antioxidant capacity is subsequently replenished by glutathione reductase, catalyzed by riboflavin, with NADPH as a proton donor. Thus, the intracellular GSH level reflects a dynamic balance between rates of GSH synthesis, consumption, regeneration and degradation. Indeed, GSH turnover is extremely rapid, ranging from 50 to 100% per day in erythrocytes and to 300 to 400% per day in the liver [3]. Among other antioxidant pathways is heme oxygenase (HO)-1, which promotes the formation of bilirubin, which serves as a potent antioxidant. As detailed bellow, HO-1 induction during sepsis might be an important mechanism in reducing oxidative stress [4].

Diverse sites of ROS generation, their stability, membrane permeability and the compartmental activities of antioxidant systems govern the intracellular distribution and concentrations of ROS. Under normal physiologic conditions, ROS exist at low levels in intracellular compartments, and even play a role in the physiologic regulation of tubular functions and renal microcirculation.

Enhanced Oxidative Stress: Mechanisms and Outcome

Profound cellular hypoxia, as occurs in various forms of AKI, may lead to the disruption of the mitochondrial electron transport system, and in different cellular compartments to the disengagement of iron and copper in redox-sensitive compounds from their protein envelopes. These changes amplify ROS generation and in particular may catalyze the formation of highly toxic hydroxyl-radical from hydrogen peroxide [5]. Importantly, in this perspective, iron excess amplifies ROS generation, inflammation and renal dysfunction, as clearly shown in a rodent sepsis model [6]. ROS attack cellular compounds, including proteins, nucleic acids and plasma membrane components. The latter outcome leads to membrane injury and leakage of mitochondrial superoxide with further cellular damage. ROS also increase vascular permeability and platelet adhesion, cause endothelial dysfunction, and enhance tubular transport and oxygen consumption [7]. Medullary oxygenation may be particularly hampered by ROS, as vasa recta are highly sensitive to ROS generation [8]. Thus, the oxidative stress-induced combined alterations of regional microcirculation and enhanced oxygen demand for transport activity may generate a feed-forward loop of renal parenchymal hypoxia-reoxygenation, linked with ROS-mediated injury. Finally, ROS may interfere with renal adaptation to stress: enhanced cellular ROS concentrations, as occurs with shock and tissue hypoxia, increase the activity of prolyl hydroxylases that degrade hypoxia-inducible factor (HIF)-α. This effect, which can be blocked by ROS scavengers, may lead to HIF destabilization and inadequate cellular HIF response to hypoxic stress. For example, we found that hypoxia-mediated HIF expression in the diabetic renal medulla is substantially improved by the administration of the membrane-permeable superoxide dismutase mimetic tempol. It is therefore tempting to assume that ROS scavengers might improve tissue adaptive responses to hypoxia, coupled with oxidative stress. However, contradicting evidence exists, indicating that ROS might trigger HIF in the absence of hypoxia. Thus, the impact of oxidative stress upon HIF-mediated cellular responses needs further assessment.

Table 1. Modes of detection of oxidative stress in vivo

Determination of oxidation products
Proteins
Thiol oxidation
Carbonyl formation
Advanced oxidation protein products
Amino acids
3-nitrotyrosine
Membrane lipids
Malondialdehyde
Lipid hydroxyperoxides
Arachidonic acid derivatives
F_2 isoprostanes
DNA
8-hydroxy-2′-deoxyguanosine
Determination of diminished antioxidant capacity
GSH content in blood and tissue samples
Cysteine/cystine ratio, or homocysteine/homocystine ratio in blood samples
Superoxide-dismutase and catalase activity in tissue samples

Evidence for Enhanced Renal ROS Generation in Sepsis

Direct determination of ROS, such as superoxide, hydrogen peroxide, hydroxyl radical and peroxynitrite is difficult due to their short half-life and compartmental distribution (i.e. intramitochondrial, cytoplasmic or extracellular). Therefore, techniques of real-time direct detection of ROS are limited to in vitro settings. Indeed, Huet et al. [9], using 2,7-dichlorodihydrofluorescein diacetate for ROS detection, reported that cultured human venous endothelial cells incubated in plasma of patients with septic shock display enhanced cellular ROS formation, reduced cellular GSH and increased cell death rate – parameters ameliorated by the addition of GSH. They suggest that inflammatory cytokines in the plasma of septic patients, such as TNF-α might be responsible for the enhanced cellular ROS production. This implies that during sepsis, ROS formation within the kidneys and renal vasculature is likely affected by a systemic inflammatory response, and is not necessarily a solely local phenomenon.

As outlined in table 1, oxidative stress in vivo is usually conveniently assessed by the determination of oxidized plasma and cellular components, such as lipids, arachidonic acid derivatives, proteins, amino acids or DNA. Furthermore, oxidative stress can be extrapolated from declining thiol antioxidants, such as GSH, and the ratio of cysteine/cystine or homocysteine/homocystine.

Enhanced ROS generation in sepsis has been documented in numerous experimental and clinical settings, though evidence for a local renal oxidative stress is more limited. Koksal et al. [10] reported that in rats subjected to cecal ligation and perforation (CLP), a commonly used sepsis model, malondialdehyde (MDA) was substantially increased in kidneys, lungs and liver, with a reciprocating fall in GSH content in these tissues. Conceivably, this reflects progressive oxidative stress, manifested by lipid peroxidation with consumption and depletion of thiol antioxidant capacity.

In a rat sepsis model induced by peritoneal seeding with *Escherichia coli*, declining kidney function was associated with increased cortical and medullary MDA content. Furthermore, renal dysfunction correlated with reduced catalase and superoxide-dismutase activity, suggesting that altered antioxidant capacity might contribute to sepsis-induced ROS-mediated AKI [11].

Human data regarding ROS generation in AKI in general and in septic patients is also quite compelling. Himmelfarb et al. [12] report that in critically ill patients with AKI, plasma protein thiols fell nearly 60% as compared with control healthy subjects, whereas plasma protein carbonyl content increased tenfold, much more than in critically ill patients without AKI or in patients with end-stage kidney disease. Thus, irrespective to its cause, AKI in general is associated with oxidative stress. Sepsis-induced AKI is specifically associated with enhanced ROS generation; as reported by Silva et al. [13], advanced oxidation protein products doubled in the plasma of septic AKI patients, as compared with control subjects, or with septic patients without AKI. Furthermore, there was a trend of higher advanced oxidation protein products levels in those patients needing renal replacement therapy, though this difference fell short of statistical significance.

Likewise, Lyons [14], studying pediatric patients, found that GSH consumption increased in septic individuals and that the GSH synthetic rate declined, culminating in reduced whole blood GSH. Cheng et al. [15] found that urinary 8-hydroxy-2'-deoxyguanosine, a marker of DNA breakdown and oxidation, is enhanced in septic patients, predicting their clinical severity and outcomes. Nevertheless, its origin might be systemic and not necessarily reflect enhanced local production within the urinary system.

Glutathione Kinetics during Sepsis

As stated above, GSH consumption and depletion develops during early experimental sepsis, though its hepatic and muscle production initially increases six- and twofold, respectively [3, 14]. However, as sepsis persists, GSH depletion intensifies, as its regeneration declines due to substrate and cofactor depletion (glutamine, cysteine, selenium), direct hormonal (glucagon, catecholamines) and cytokine-induced inhibitory effects (erythropoietin, TGF-β), and decaying stimulatory factors such as thyroid hormones and gonadotropins [3].

Fläring [16], exploring GSH kinetics during *early* endotoxemia reported that in 8 healthy volunteers infused with lipopolysaccharide, plasma GSH gradually declined over 4 h, while the whole blood GSH fractional synthesis rate tended to increase. This data suggests a ROS effect, i.e. GSH consumption and enhanced synthesis; however, whole blood and muscle total GSH and GSH redox state were not altered and liver GSH production was not determined.

Origin and Mechanisms of Sepsis-Induced Reactive Oxygen Species Formation

Sepsis leads to systemic inflammatory response with neutrophil infiltration and activation. This provokes a phagocyte-associated respiratory burst via the activation of membrane-bound NADPH oxidase. Furthermore, renal parenchymal myeloperoxidase activity markedly increases as well, as shown in a murine CLP sepsis model [17]. Myeloperoxidase catalyzes the production of hypochlorous acid, another potent oxidant. Membrane-bound (NADPH) oxidases are also expressed in neutrophils and endothelial cells, and catalyze the production of superoxide [18]. In sum, oxidative stress originates in inflammatory cells, as has clearly been documented in septic patients with systemic inflammatory response and multiorgan dysfunction [19].

Sepsis also leads to the induction of inducible NOS (iNOS) with exponentially increasing generation of nitric oxide. This promotes the formation of the highly toxic peroxynitrite following interaction with superoxide. Importantly, evolving nitrosative stress also leads to mitochondrial complex I S-nitrosylation and consequent dysfunction, particularly when the mitochondrial GSH pool is oxidized. As a result, superoxide production from complex I is augmented with a feed-forward loop of ROS generation [20].

Profoundly enhanced xanthine oxidase activity was also noted in pediatric septic patients, especially in the earlier phases of this disorder, which parallel the decline in the antioxidant capacity, underscoring its additive role in the oxidative stress buildup during sepsis [21]. Furthermore, enhanced xanthine oxidase activity paralleled the APACHE II score and predicted mortality in septic patients [22].

Endotoxemia causes systemic vasodilation, related to a large extent to enhanced nitric oxide production via iNOS induction, and to vascular adrenoreceptor hyporeactivity. Shock, myocardial depression and renal vasoconstriction might lead to reduced renal parenchymal oxygenation [23] with subsequently enhanced local ROS formation, as discussed above. Indeed Wu et al. [24], using intravital videomicroscopy in a CLP mice model, found that sepsis reduced cortical peritubular capillary blood flow, associated with local generation of ROS in ischemic tubular cells (indicated by DHR oxidation to rhodamine fluorescence, and by nitrotyrosine immunostaining). Furthermore, Johannes et al. [25]

Facing this complex disease entity, alleviation of oxidative stress single-handedly is unlikely to be effective in the prevention of sepsis-associated renal dysfunction. However, the addition of antioxidants to a comprehensive treatment strategy seems a reasonable approach. Additional confirmatory experimental studies using antioxidants, iron-chelating agents and selective iNOS inhibitors in large animals might enable their introduction in clinical trials. Future therapeutic avenues should also include the development of better antioxidants with improved cellular and mitochondrial penetration.

Acknowledgements

This report is supported by the Israel Science Foundation (grant No. 1473/08) and the Harvard Medical Faculty Physicians at Beth Israel Deaconess Medical Center, Boston, Mass., USA.

References

1 Koyner JL, Sher Ali R, Murray PT: Antioxidants: Do they have a place in the prevention or therapy of acute kidney injury? Nephron Exp Nephrol 2008;109:e109–e117.
2 Pinsky MR: Sepsis and multiple organ failure; in Ronco C, Bellomo R, Kellum JA (eds): Acute Kidney Injury. Contrib Nephrol. Basel, Karger, 2007;156:47–63.
3 Biolo G, Antonione R, De Cicco M: Glutathione metabolism in sepsis. Crit Care Med 2007;35(9 Suppl):S591–S595.
4 Jao HC, Lin YT, Tsai LY, Wang CC, Liu HW, Hsu C: Early expression of heme oxygenase-1 in leukocytes correlates negatively with oxidative stress and predicts hepatic and renal dysfunction at late stage of sepsis. Shock 2005;23:464–469.
5 Legrand M, Mik EG, Johannes T, Payen D, Ince C: Renal hypoxia and dysoxia after reperfusion of the ischemic kidney. Mol Med 2008;14:502–516.
6 Zager RA, Johnson AC, Hanson SY: Parenteral iron therapy exacerbates experimental sepsis. Kidney Int 2004;65: 2108–2112.
7 Juncos R, Garvin JL: Superoxide enhances Na-K-2Cl cotransporter activity in the thick ascending limb. Am J Physiol 2005;288:F982–F987.
8 Cao C, Edwards A, Sendeski M, Lee-Kwon W, Cui L, Cai CU, Patzak A, Pallone TL: Intrinsic nitric oxide and superoxide production regulates descending vasa recta contraction. Am J Physiol Renal Physiol 2010;299:F1056–F1064.
9 Huet O, Cherreau C, Nicco C, Dupic L, Conti M, Borderie D, Pene F, Vicaut E, Benhamou D, Mira JP, Duranteau J, Batteux F: Pivotal role of glutathione depletion in plasma-induced endothelial oxidative stress during sepsis. Crit Care Med 2008;36: 2328–2334.
10 Koksal GM, Sayilgan C, Aydin S, Oz H, Uzun H: Correlation of plasma and tissue oxidative stresses in intra-abdominal sepsis. J Surg Res 2004;122:180–183.
11 Edremitlioglu M, Kilic D, Oter S, Kisa U, Korkmaz A, Coskun O, Bedir O: The effect of hyperbaric oxygen treatment on the renal functions in septic rats: relation to oxidative damage. Surg Today 2005;35:653–661.
12 Himmelfarb J, McMonagle E, Freedman S, Klenzak J, McMenalmin E, Le P, Pupim LB, Ikizler TA, The PICARD Group: Oxidative stress is increased in critically ill patients with acute renal failure. J Am Soc Nephrol 2004;15:2449–2456.

13 Silva S, de Cal M, Cruz D, Lentini P, Corradi V, Gallo G, Salvatori G,Verbine A, Pogoshyan L, Nalesso F, Brendolan A, Piccinni P, Ronco C: Oxidative stress and monocyte reprogramming in septic patients with acute kidney injury requiring CRRT. Blood Purif 2008;26:188–192.

14 Lyons J, Rauh-Pfeiffer A, Ming-Yu Y, Lu XM, Zurakowski D, Curley M, Collier S, Duggan C, Nurko S, Thompson J, Ajami A, Borgonha S, Young VR, Castillo L: Cysteine metabolism and whole blood glutathione synthesis in septic pediatric patients. Crit Care Med 2001;29:870–877.

15 Cheng WE, Shih CM, Hang LW, Wu KY, Yang HL, Hsu WH, Hsia TC: Urinary biomarker of oxidative stress correlating with outcome in critically septic patients. Intensive Care Med 2007;33: 1187–1190.

16 Fläring UB, Hebert C, Wernerman J, Hammarqvist F, Rooyackers OE: Circulating and muscle glutathione turnover in human endotoxaemia. Clin Sci (Lond) 2009;117:313–319.

17 Ritter C, Andrades ME, Reinke A, Menna-Barreto S, Moreira JC, Dal-Pizzol F: Treatment with N-acetylcysteine plus deferoxamine protects rats against oxidative stress and improves survival in sepsis. Crit Care Med 2004;32:342–349.

18 Jaber BL, Pereira BJC, Bonventre JV, Balakrishnan VS: Polymorphism of host response genes: implications in the pathogenesis and treatment of acute renal failure. Kidney Int 2005;67:14–33.

19 Tanjoh K, Shima A, Aida M, Tomita R, Kurosu Y: Nitric oxide and active oxygen species in severe sepsis and surgically stressed patients. Surg Today 1995;25:774–777.

20 Dahm CC, Moore K, Murphy MP: Persistent S-nitrosation of complex I and other mitochondrial membrane proteins by S-nitrosothiols but not nitric oxide or peroxynitrite. Implications for the interaction of nitric oxide with mitochondria. J Biol Chem 2006;281:10056–10065.

21 Németh I, Boda D: Xanthine oxidase activity and blood glutathione redox ratio in infants and children with septic shock syndrome. Intensive Care Med 2001;27:216–221.

22 Luchtemberg MN, Petronilho F, Constantino L, Gelain DP, Andrades M, Ritter C, Moreira JC, Streck EL, Dal-Pizzol F: Xanthine oxidase activity in patients with sepsis. Clin Biochem 2008;41:1186–1890.

23 Johannes T, Ince C, Klingel K, Unertl KE, Mik EG: Iloprost preserves renal oxygenation and restores kidney function in endotoxemia-related acute renal failure in the rat. Crit Care Med 2009;37: 1423–1432.

24 Wu L, Gokden N, Mayeux PR: Evidence for the role of reactive nitrogen species in polymicrobial sepsis-induced renal peritubular capillary dysfunction and tubular injury. J Am Soc Nephrol 2007;18:1807–1815.

25 Johannes T, Mik EG, Ince C: Nonresuscitated endotoxemia induces microcirculatory hypoxic areas in the renal cortex in the rat. Shock 2009, 31:97–103.

26 Sprong RC, Winkelhuyzen-Janssen AML, Aarsman CJ, van Oirschot JF, van der Bruggen T, van Asbeck BS: Low-dose N-acetylcysteine protects rats against endotoxin-mediated oxidative stress, but high-dose increases mortality. Am J Respir Crit Care Med 1998;157:1283–1293.

27 Villa P, Ghezzi P: Effect of N-acetyl-L-cysteine on sepsis in mice. Eur J Pharmacol 1995;292:341–344.

28 Spapen HD, Diltoer MW, Nguyen DN, Hendrickx I, Huyghens LP: Effects of N-acetylcysteine on microalbuminuria and organ failure in acute severe sepsis: results of a pilot study. Chest 2005;127:1413–1419.

29 Molnar Z, Szakmany T, Koszegi T: Prophylactic N-acetylcysteine decreases serum CRP but not PCT levels and microalbuminuria following major abdominal surgery. A prospective, randomised, double-blinded, placebo-controlled clinical trial. Intensive Care Med 2003;29:749–755.

30 Schaller G, Pleiner J, Mittermayer F, Posch M, Kapiotis S, Wolzt M: Effects of N-acetylcysteine against systemic and renal hemodynamic effects of endotoxin in healthy humans. Crit Care Med 2007;35:1869–1875.
31 Senoglu N, Yuzbasioglu MF, Aral M, Ezberci M, Kurutas EB, Bulbuloglu E, Ezberci F, Oksuz H, Ciragil P: Protective effects of N-acetylcysteine and beta-glucan pretreatment on oxidative stress in cecal ligation and puncture model of sepsis. J Invest Surg 2008;21:237–243.
32 Matejovic M, Krouzecky A, Martinkova V, Rokyta R, Kralova H, Treska V, Radermacher P, Novak I: Selective inducible nitric oxide synthase inhibition during long-term hyperdynamic porcine bacteremia. Shock 2004;21:458–465.
33 Dare AJ, Phillips AR, Hickey AJ, Mittal A, Loveday B, Thompson N, Windsor JA: A systematic review of experimental treatments for mitochondrial dysfunction in sepsis and multiple organ dysfunction syndrome. Free Radic Biol Med 2009;47:1517–1525.

Prof. Samuel N. Heyman
Dept. of Medicine, Hadassah Hospital, Mt. Scopus
P.O. Box 24035
Jerusalem 91240 (Israel)
Tel. +1 972 2 584 4526, E-Mail Heyman@cc.huji.ac.il

Kellum JA, Ronco C, Vincent J-L (eds): Controversies in Acute Kidney Injury.
Contrib Nephrol. Basel, Karger, 2011, vol 174, pp 149–155

Pathophysiology of Acute Kidney Injury to Chronic Kidney Disease: Maladaptive Repair

Li Yang[a,c] · Benjamin D. Humphreys[a,b] · Joseph V. Bonventre[a,b]

[a]Renal Division, Brigham and Women's Hospital and Department of Medicine, Harvard Medical School, Boston, Mass., and the [b]Harvard Stem Cell Institute, Cambridge, Mass., USA; [c]Renal Division, Beijing University First Hospital, Beijing, China

Abstract

After damage the kidney has the ability to repair itself. With mild injury this repair can result in the return to a structural and functional state that is indistinguishable from normal. However, when the repair is more severe or is superimposed on baseline kidney abnormalities, the repair process can lead to fibrosis, which can facilitate progression to chronic kidney disease. Epidemiological studies now show that patients who have had acute kidney injury have a marked increase in their risk for the development of end-stage renal disease. Recent data have redefined the role of the surviving epithelial cells in fibrosis and attribute myofibroblast expansion to perivascular and interstitial fibroblasts. After severe injury, the proximal tubule cellular response is impaired with its proliferative response altered due to cell cycle arrest at the G2/M phase of the cell cycle, resulting in generation of profibrotic factors including cytokines, growth factors and matrix proteins.

Whether AKI is associated with ischemia reperfusion injury, sepsis or toxins, there is a rapid loss of proximal tubular cell cytoskeletal integrity and cell polarity. There is shedding of the proximal tubule brush border, loss of polarity with mislocalization of adhesion molecules and other membrane proteins such as the Na+K+ATPase and β-integrins [1, 2]. Normal cell-cell interactions are disrupted with injury. When the injury is severe there is apoptosis and necrosis [2]. Viable and nonviable cells are desquamated leaving regions where the basement membrane remains the only barrier between the filtrate and the peritubular interstitium. Although the kidney can recover from an injury, this recovery process can often be suboptimal in animals and humans, especially when the injury is severe or on a background of underlying chronic kidney disease (CKD). Acute kidney injury (AKI) can thus result in incomplete repair and persistent tubulointerstitial inflammation, with proliferation of fibroblasts and excessive deposition of extracellular matrix, a common feature of many different kinds of kidney diseases and a primary determinant of progression to end-stage renal failure [3, 4]. Tubulointerstitial fibrosis is a hallmark of CKD and it is often much more severe than the glomerular pathology that may have been integral to the initiation of the renal disease. Fibrosis is a characteristic of the maladaptive repair that accompanies severe AKI alone or AKI in the setting of underlying kidney disease [5].

It has been proposed that the epithelial cell may transform into a fibroblast and contribute to the fibrosis which is a hallmark of CKD. Recent studies in animals, however, have challenged this concept and have provided new insight into the mechanisms responsible for normal and abnormal repair and draw an important link between injury, abnormal repair and the development of fibrosis. This propensity to develop fibrosis after severe injury in animals provides a pathophysiological underpinning to the progression of CKD that is often seen in humans who sustain AKI in the setting of underlying CKD.

In this short review we will discuss, first, the effects of AKI on the progression of CKD and development of end-stage renal disease in humans and then discuss the pathophysiological factors which link severe injury to maladaptive repair of the kidney after injury.

Epidemiological Studies

In a large community-based cohort of patients with CKD in northern California with a baseline estimated glomerular filtration rate (eGFR) of less than 45 ml/min per 1.73 mm^2, there was a high risk of developing end-stage renal disease within 30 days of hospitalization if the patients developed AKI during the hospitalization: 42 and 63% if baseline eGFR was between 30–44 or 15–29 ml/min per 1.73 m^2, respectively [6]. In another cohort of hospitalized older patients 67 years of age or older, the adjusted hazard ratio of development of end-stage renal disease was 42.2 for patients who developed AKI in the context of CKD relative to those without renal disease, 13.0 for patients who developed AKI without CKD and 8.4 for patients with CKD without AKI [7]. Other studies have shown that patients who develop AKI may have long-term adverse kidney outcomes even if they did not have kidney disease at the time of development of AKI [8].

Pathophysiology of Chronic Kidney Disease After Acute Kidney Injury

Many factors contribute to the CKD that develops after acute injury. A hallmark of CKD is fibrosis and loss of functional nephrons. AKI is an inflammatory disease [9] and the components contributing to that inflammation can persist beyond the initial insult. The cellular infiltrate in kidney after ischemia/reperfusion injury contributes to both parenchymal damage and also provides a protective response. As will be discussed below, different cell populations can make different types of contributions.

Another important feature of abnormal recovery from AKI is the loss of tissue capillaries. The number of microvessels in the kidney declines, potentially facilitated by the downregulation of angiogenic factors such as the VEGF pathway [10] and upregulation of angiogenesis inhibitors. A reduced number of vessels is associated with chronic hypoxia which can trigger the proinflammatory responses and cellular infiltrate leading to increased tubulointerstitial fibrosis which can be relentless since the activating force, hypoxia, will be persistent or get worse over time. Small vessels that survive may have blood flow compromised by endothelial cell swelling. Components of the inflammatory response, including monocytes and T lymphocytes, may promote neovascularization through production of proangiogenic factors although other components of the inflammatory response such as T regulatory cells, which have been shown to inhibit inflammation [11], also inhibit angiogenesis in a hindlimb ischemia model [12].

Inflammatory Effector Cells

Immune/inflammatory cells can potentiate renal injury by generating mediators including cytokines, chemokines, eicosanoids and reactive oxygen species; recruit leukocytes; and upregulate adhesion molecules that engage counter-receptors on the activated endothelium. These cytokines also disrupt cell-matrix adhesion dependent on β_1-integrin, inducing cell shedding into the lumen. Neutrophils, monocytes/macrophages dendritic cells and T-lymphocytes all play important roles in the inflammatory response to ischemia/reperfusion.

In later phases after kidney injury, macrophages and T lymphocytes predominate over neutrophils. Delayed mononuclear cell infiltration is believed to regulate the outcome of the repair process, modulating tissue remodeling. Macrophages represent a heterogeneous population of monocytic cells that contribute not only to kidney fibrosis but also to beneficial effects on repair and matrix remodeling [13]. F4/80+ macrophages in the kidney increase at 1 h after reperfusion, peak at 24 h and persist for 7 days [14]. Dendritic cells are also major effector cells linking injury to immune response in the kidney [15]. During acute tubular injury, these resident myeloid cells are activated and additional blood monocytes are recruited to the tissue. Dendritic cells and macrophages are key contributors to the innate immune response associated with ischemia/reperfusion contributing to the inflammatory response by releasing soluble mediators and directly activating other cells [16]. Dendritic cells can activate naïve T cells, thus linking the innate immune response to adaptive immunity [17]. T cells contribute to the injury response, but likely also play a role in the repair phase of

AKI. TCR-β(+) CD4(+) CD25(+)Foxp3(+) regulatory T cells (T_{reg}) infiltrate the kidney after 3 or 10 days in the mouse model of ischemia [11]. T_{reg} cell depletion resulted in worse functional injury and increased mortality.

Monocytes contribute to fibroblast proliferation and the development of fibrosis [18]. Important contributors to the response to injury are the factors that are responsible for the recruitment of different populations of monocytes which may determine the ultimate fate of the tissue after injury: complete repair with little sequelae versus fibrosis with a tendency to more or less progress. Endothelial cell activation with ischemia and reperfusion can be a proximal mediator of the inflammatory response. It is likely that tubular cells themselves also regulate this inflammatory response although distinguishing effects of tubular cells from those of endothelial cells is not always straightforward given that most etiologies of injury have activating effects on both the vessels and the tubules. In addition, the loss of capillaries that characterize ischemia/reperfusion injury [19] can set up a positive feedback process by which the loss of vessels results in hypoxia to areas of the nephron which then generate hypoxia-inducible factors. This can add to the inflammatory response and the production of vasoconstrictive agents can further reduce blood flow, leading to more capillary rarefaction and ultimately nephron loss.

Repair of the Epithelium after Severe Ischemic or Toxic Injury

The predominant mechanism of normal repair after ischemic tubular injury in the adult mammalian kidney involves regeneration by surviving tubular epithelial cells [20, 21]. Repair can either leave no lasting evidence of damage or can result in fibrotic lesions which can result in progressive organ dysfunction. While much has been written about the potential role of epithelial-to-mesenchymal transition in this process of fibrosis, Humphreys et al. [21, 22], based on lineage tracing experiments in the mouse, established that epithelial cells do not contribute to the generation of myofibroblasts directly by transdifferentiation in vivo. Rather the myofibroblasts are generated from fibroblasts, many of which are perivascular fibroblasts, or pericytes.

Although the epithelial cell does not transdifferentiate into a myofibroblast, the damaged proximal epithelial cell can nevertheless orchestrate the process of fibrosis by generating profibrogenic cytokines. Using various models of AKI, including bilateral and unilateral ischemia/reperfusion, aristolochic acid nephropathy, and unilateral ureteral obstruction, we have mapped cell cycle progression in vivo and found large differences among injury models resulting in 'normal' nonfibrotic repair and those resulting in 'fibrotic' repair [23]. We have found a causal association between epithelial cell cycle arrest at G2/M and the development of fibrosis. This required the establishment of new protocols to identify various stages of the cell cycle in vivo (fig. 1). G2/M-arrested proximal

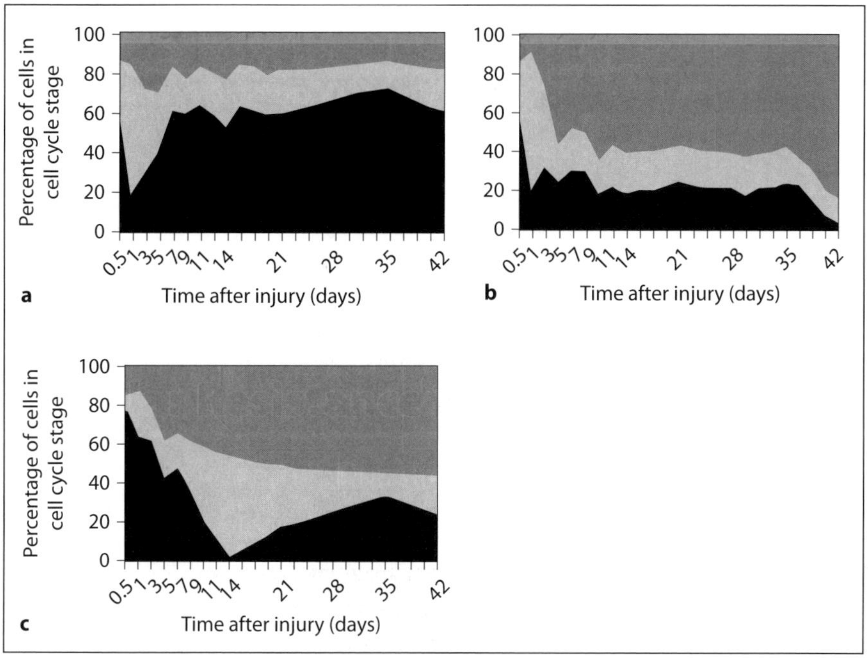

Fig. 1. Cell cycle distribution (G1, S and G2/M) of tubular cells in moderate ischemia/reperfusion injury (**a**), unilateral ischemia/reperfusion injury (**b**) and aristolochic acid nephropathy (**c**) models as a function of time after the insult. To analyze the cell cycle distribution of proximal tubular epithelial cells in vivo, proliferating cells (G1, S, G2 and M phase) were labeled in vivo with K_i-67-specific antibody, identified in the S phase by the uptake of BrdU, detected by staining with antibody to BrdU and detected in the G2/M phase by quantifying the number of tubular cells stained with an antibody to phosphohistone H3 at Ser10 (p-H3). Reproduced with permission from Yang et al. [23].

tubular epithelial cells activate the c-jun NH_2-terminal kinase (JNK) signaling cascade that acts to upregulate profibrotic cytokines production. Importantly, the fibrotic fate of the injured kidney can be rescued either by using a JNK inhibitor, or bypassing the G2/M arrest by administration of a p53 inhibitor or the removal of the contralateral kidney, providing new therapeutic approaches. This introduces a new dimension to our understanding of the development of fibrosis after injury and provides insight into our recent clinical recognition that AKI is an important potentiating factor in the progression of CKD [7].

Conclusions

The processes involved with the repair of the kidney after injury are complex. It is of great clinical significance that the repair process often leads to worsening of CKD and greatly enhances the chance that the individual will progress to end-stage renal disease. Inflammation plays an important role in the regulation of this maladaptive reparative response. We have learned a great deal recently about the roles of innate and acquired immunity in the injury phase and processes related to repair of the epithelial layer. Recent data challenge the paradigm that epithelial-to-mesenchymal transformation is a primary contributor to the fibrotic response. After severe injury, the epithelial proliferative response is impaired with cell cycle arrest at the G2/M phase of the cell cycle leading to a profibrotic secretory phenotype resulting in the generation of cytokines, growth factors and matrix molecules which result in the proliferation of perivascular fibroblasts and interstitial fibrosis, important components of the pathophysiology of CKD. Collectively, these advances in our understanding of maladaptive repair have identified new therapeutic targets to promote beneficial repair and reduce fibrosis.

Acknowledgements

This work was supported by the National Institutes of Health (grants DK39773, DK72381, DK54741). L.Y. was supported by a fellowship from the International Society of Nephrology.

References

1 Zuk A, Bonventre JV, Brown D, Matlin KS: Polarity, integrin, and extracellular matrix dynamics in the postischemic rat kidney. Am J Physiol 1998;275:C711–C731.

2 Thadhani R, Pascual M, Bonventre JV: Acute renal failure. New Engl J Med 1996;334:1448–1460.

3 Forbes JM, Hewitson TD, Becker GJ, Jones CL: Ischemic acute renal failure: long-term histology of cell and matrix changes in the rat. Kidney Int 2000;57:2375–2385.

4 Macedo E, Bouchard J, Mehta RL: Renal recovery following acute kidney injury. Curr Opin Crit Care 2008;14:660–665.

5 Venkatachalam MA, Griffin KA, Lan R, Geng H, Saikumar P, Bidani AK: Acute kidney injury: a springboard for progression in chronic kidney disease. Am J Physiol Renal Physiol 2010, E-pub ahead of print.

6 Hsu CY, Chertow GM, McCulloch CE, Fan D, Ordonez JD, Go AS: Nonrecovery of kidney function and death after acute on chronic renal failure. Clin J Am Soc Nephrol 2009;4:891–898.

7 Ishani A, Xue JL, Himmelfarb J, Eggers PW, Kimmel PL, Molitoris BA, Collins AJ: Acute kidney injury increases risk of ESRD among elderly. J Am Soc Nephrol 2009;20:223–228.

8 Coca SG, Yusuf B, Shlipak MG, Garg AX, Parikh CR: Long-term risk of mortality and other adverse outcomes after acute kidney injury: a systematic review and meta-analysis. Am J Kidney Dis 2009;53:961–973.

9 Bonventre JV, Zuk A: Ischemic acute renal failure: an inflammatory disease? Kidney Int 2004;66:480–485.

10 Basile DP, Fredrich K, Chelladurai B, Leonard EC, Parrish AR: Renal ischemia reperfusion inhibits VEGF expression and induces ADAMTS-1, a novel VEGF inhibitor. Am J Physiol Renal Physiol 2008;294:F928–F936.

11 Gandolfo MT, Jang HR, Bagnasco SM, Ko GJ, Agreda P, Satpute SR, Crow MT, King LS, Rabb H: Foxp3+ regulatory T cells participate in repair of ischemic acute kidney injury. Kidney Int 2009;76:717–729.
12 Zouggari Y, Ait-Oufella H, Waeckel L, Vilar J, Loinard C, Cochain C, Recalde A, Duriez M, Levy BI, Lutgens E, Mallat Z, Silvestre JS: Regulatory T cells modulate postischemic neovascularization. Circulation 2009;120:1415–1425.
13 Ricardo SD, van Goor H, Eddy AA: Macrophage diversity in renal injury and repair. J Clin Invest 2008;118:3522–3530.
14 Li L, Huang L, Sung SS, Vergis AL, Rosin DL, Rose CE Jr, Lobo PI, Okusa MD: The chemokine receptors CCR2 and CX3CR1 mediate monocyte/macrophage trafficking in kidney ischemia-reperfusion injury. Kidney Int 2008;74:1526–1537.
15 John R, Nelson PJ: Dendritic cells in the kidney. J Am Soc Nephrol 2007;18:2628–2635.
16 Li L, Okusa MD: Macrophages, dendritic cells, and kidney ischemia-reperfusion injury. Semin Nephrol 2010;30:268–277.
17 Reis e Sousa C: Dendritic cells in a mature age. Nat Rev Immunol 2006;6:476–483.
18 Vernon MA, Mylonas KJ, Hughes J: Macrophages and renal fibrosis. Semin Nephrol 2010;30:302–317.
19 Basile DP: The endothelial cell in ischemic acute kidney injury: implications for acute and chronic function. Kidney Int 2007;72:151–156.
20 Bonventre JV: Dedifferentiation and proliferation of surviving epithelial cells in acute renal failure. J Am Soc Nephrol 2003;14(Suppl 1):S55–S61.
21 Humphreys BD, Valerius MT, Kobayashi A, Mugford JW, Soeung S, Duffield JS, McMahon AP, Bonventre JV: Intrinsic epithelial cells repair the kidney after injury. Cell Stem Cell 2008;2:284–291.
22 Humphreys BD, Lin SL, Kobayashi A, Hudson TE, Nowlin BT, Bonventre JV, Valerius MT, McMahon AP, Duffield JS: Fate tracing reveals the pericyte and not epithelial origin of myofibroblasts in kidney fibrosis. Am J Pathol 2010;176:85–97.
23 Yang L, Besschetnova TY, Brooks CR, Shah JV, Bonventre JV: Epithelial cell cycle arrest in G2/M mediates kidney fibrosis after injury. Nat Med 2010;16:535–543, 1p following 143.

Joseph V. Bonventre, MD, PhD
Brigham and Women's Hospital
Renal Division, MRB-4
75 Shattuck Street, Boston, MA 02115 (USA)
Tel. +1 617 525 5960, E-Mail joseph_bonventre@hms.harvard.edu

Kellum JA, Ronco C, Vincent J-L (eds): Controversies in Acute Kidney Injury.
Contrib Nephrol. Basel, Karger, 2011, vol 174, pp 156–162

Models of Preventable Disease: Contrast-Induced Nephropathy and Cardiac Surgery-Associated Acute Kidney Injury

Andrew Shaw

Department of Anesthesiology, Duke University Medical Center, Durham, N.C., USA

Abstract

In order to prevent a disease, its temporal nature (or at least when it starts) needs to be clearly defined. In acute kidney injury (AKI), this is usually not possible because the current diagnostic criteria are retrospective. Contrast-induced nephropathy (CIN) and cardiac surgery-associated acute kidney injury (CSA-AKI) are both thought of as potentially preventable acute renal lesions because the timing of the insult is known precisely. While this is true, knowledge of the exact biological processes that give rise to each condition is lacking, although there are several common themes – notably ischemia – that pervade the literature describing these diseases. Despite this, progress in prevention has been slow, and to date there are no therapies indicated for preventing either CIN or CSA-AKI. The best we can currently do is to recommend aggressive parenteral hydration, avoid compounds we know are nephrotoxic, and avoid unnecessary hypoxia and hypotension. While there is still clearly a long way to go before either of these acute kidney conditions can be described as preventable, the use of major adverse kidney events – death, dialysis and incident or progressive chronic kidney disease at 90 days – as a composite endpoint in clinical trials of putative prevention agents would represent a significant step forwards.

Prevention Trial Design

In clinical trial terms, identifying and defining a primary endpoint for the study are fundamentally important parts of the planning process. In acute kidney injury (AKI), the diagnostic criteria are evolving continually and may well change in the next few years as the place of novel biomarkers of acute intrinsic renal damage and function becomes clearer. However, there is a clear unmet medical need in AKI today, and thus we cannot wait and see which definitions for clinical trials the field ultimately accepts. Rather, we need to design trials with endpoints that are robust to changes in acute terminology. This is difficult, but possible if we focus on outcomes that have relevance for all stakeholders: namely endpoints that affect quantity and quality of life, and endpoints that lead to outcomes with significance for long-term functioning.

Composite Acute Kidney Injury Prevention Trial Endpoints: Major Adverse Kidney Events

A composite endpoint is attractive in prevention trials because it allows the inclusion of several important outcomes, which provides a higher event rate, and thus makes it easier to find signal in an otherwise noisy diagnostic space. One possibility is to use the term 'major adverse kidney events', and include measures such as death, new dialysis, and incident and progressive chronic kidney disease. Death is clearly important, but the competing risk problem needs to be fully articulated. New renal replacement therapy is also important as it clearly is associated with adverse outcomes, but the decision to dialyze is not uniform. Hence an adjudicated outcome (e.g. hyperkalemia, pulmonary edema or severe metabolic acidosis) allows patients to reach the endpoint but preserves the autonomy of the treating clinician to intervene later if he or she believes it is in the patient's best interest. Incident or progressive chronic kidney disease, determined by measurement of glomerular filtration rate at 90 days, allows detection of ongoing impairment of function and captures those with impaired function at baseline.

Contrast-Induced Nephropathy and Cardiac Surgery-Associated Acute Kidney Injury as Preventable Conditions

Contrast-induced nephropathy (CIN) and cardiac surgery-associated acute kidney injury (CSA-AKI) share many characteristics. In most situations the exact timing (and sometimes the magnitude) of the insult is known precisely, and the populations of patients in which they occur overlap significantly. Mechanistically, it is believed there are similarities (ischemia, apoptosis) and differences (atheroembolism, pigment nephropathy) between the two conditions. However, it is the temporal nature of the insult that makes these two common acute renal syndromes potentially the most amenable to prevention. This paper reviews some of the characteristics of CIN and CSA-AKI that explain why there is so much interest in them as models of preventable disease.

Epidemiology

There is significant overlap between the population of patients that experience CIN and CSA-AKI, namely patients with correctable heart disease [1]. The incidence of CIN varies in the literature, and is highly dependent on baseline renal function [2]. Rudnick et al. [3] suggest that as a rule of thumb the incidence of CIN approximates 10 times the starting serum creatinine (e.g. in patients with a baseline creatinine of 1.9 the predicted incidence is 19%). The incidence of CSA-AKI varies according to the definition used, the operation in question, the center in which the surgery is performed and many other factors [4]. If RIFLE-R is used as a diagnostic cutoff, then the incidence approaches 25% overall, but the incidence is much lower (about 5–10%) if RIFLE-F is used. The RIFLE criteria allow identification of patients at risk for severe AKI (the incidence of RIFLE-F is far higher in patients who have already reached R) and progression to I+F (and perhaps time spent in the F stage) may therefore represent a useful endpoint for prevention trials. These criteria are based on changes in serum creatinine, and have been extensively validated in both the CIN and CSA-AKI literature [5]. It is worth mentioning that novel biomarkers of renal injury and function are now available for clinical use (e.g. urinary NGAL in Europe, and cystatin C in the USA), but it is not yet clear how these markers will fit into the diagnostic arsenal. It is likely that the diagnostic criteria for risk and diagnosis of both CIN and CSA-AKI will change in future years as the exact place of the biomarkers currently under study becomes clearer [6].

Pathophysiology

Almost every patient who undergoes cardiac surgery will have previously undergone either cardiac catheterization or CT scanning of the chest, and thus it is probably fair to say that recent exposure to radiocontrast is universal in the CSA-AKI population. It is not known the extent to which CIN affects CSA-AKI, but undoubtedly there is some interaction, particularly in those who receive large volumes of contrast in the immediate preoperative period (i.e. nonelective cases) [7].

Ischemia

Administration of radiocontrast material is known to reduce renal blood flow, both globally and also at the level of the outer medulla [8]. This is predominantly mediated via intrarenal vasoconstriction, but is also in part effected by endothelin release [9]. The already (relatively) hypoxic medulla (PO_2 10–15 mm Hg) thus becomes profoundly hypoxic and renders the local tissues susceptible to ischemic damage. In the context of intravascular volume depletion, concurrent administration of vasopressors (e.g. during a period of hypotension) and perhaps anemia, the result is an ischemic insult whose magnitude depends on the duration of tissue hypoxia [10].

Ischemia is also involved in CSA-AKI, as the combination of hypotension, anemia and hypoxia is also undoubtedly present during the conduct of routine cardiopulmonary bypass. It is not clear whether nonpulsatile circulation induces an extra burden on intrarenal hemodynamics, as a clear advantage of pulsatile cardiopulmonary bypass has never been documented, despite extensive efforts. Blood transfusion and anemia have both been shown to be detrimental in terms of risk of CSA-AKI, making the decision to transfuse a difficult one. However, most authorities recommend transfusion if on-pump hematocrit reduces to less than 22%.

Atheroembolism
One feature of CSA-AKI that is different from other causes is the presence of atheroembolic disease. When the aorta is instrumented for cannulation, some atheroembolism is inevitable, and when the atheromatous burden is severe, this is associated with end-organ compromise affecting brain, kidney and probably also the gastrointestinal tract. There have been many attempts to reduce this burden, using novel cannulae with baskets and other devices to attempt to 'catch' large fragments of plaque as they dislodge, but the use of epiaortic ultrasound for identification of the optimal cannulation site is perhaps the most useful technique in practice. This is now widely available in cardiac operating rooms [11].

Pigment Nephropathy
Of particular importance for CSA-AKI is the role of cardiopulmonary bypass-induced hemolysis. It is likely that mitigation of the renal effects of increased plasma free hemoglobin is the reason for the apparent success of infusions of sodium bicarbonate in reducing the extent of the serum creatinine rise when compared with an infusion of sodium chloride of similar 'tonic' load [12]. Alkalinization of the urine may even be a better therapeutic target than administration of a set dose of buffer to high-risk cardiac surgery patients since it allows both personalization (e.g. acetazolamide for patients with systemic alkalemia, in whom base infusions would be difficult to justify) and titration (e.g. achieve a urine pH of 7) of the treatment. A large trial of bicarbonate therapy is currently underway in Europe and Australia.

Prevention Strategies

By far the simplest prevention strategy is to avoid contrast and cardiac surgery altogether. This, however, is not possible with today's techniques for treating heart disease, and many patients with correctable heart disease are at risk for both CIN and CSA-AKI [7]. Unlike sepsis-associated AKI, the timing of the insult is known and this provides an advantage for clinicians as they prepare patients for intervention. General management measures, such as avoiding hypotension and hypoxemia, are important and complement more specific interventions such as intravenous fluid therapy. For CIN in particular this has been repeatedly demonstrated to be of benefit and is now recommended for all patients receiving contrast [3]. It is not clear whether this needs to be aggressively managed (e.g. by matching urine volume exactly with fluid administered) or whether simply avoiding dehydration is sufficient. The type of fluid is potentially important also, and studies have shown that using intravenous bicarbonate solutions may be more protective than using sodium chloride solutions.

Bicarbonate therapy has also been shown to reduce the incidence of RIFLE-defined kidney injury in the cardiac surgery setting [12], but further studies are required before urinary alkalinization is adopted as standard of care for all cardiac surgery patients. It is likely that as well as reducing the generation of oxygen free radicals, bicarbonate therapy enhances excretion of free hemoglobin (arising due to hemolysis from the cardiopulmonary bypass machine) and thus attenuates the pigment nephropathy that is a component of CSA-AKI.

As mentioned above, the role of anemia and blood transfusion is particularly relevant for CSA-AKI since both are risk factors for AKI. An on-pump hematocrit of less than 21% is generally avoided since this significantly increases the risk of kidney impairment in the postoperative period, but in patients with pre-existing anemia, this may not be possible without blood transfusion. Perfusion techniques (such as continuous rapid autologous priming) may attenuate the hemodilution somewhat, but some degree of anemia is usually unavoidable in patients who do not begin the procedure with a normal red cell mass.

In terms of specific pharmacotherapy for preventing CIN and CSA-AKI, the data are disappointing. Many agents have been tried for both conditions, but none have been proven effective. A summary of therapies can be seen in table 1.

Conclusions

CIN and CSA-AKI are common (at least theoretically) avoidable conditions that affect many patients presenting for cardiac procedures. Their occurrence adversely affects the outcome of other interventions, and thus they represent important disease states for prevention trials of novel compounds and existing techniques. Avoiding hemodynamic compromise, anemia, blood transfusion,

Table 1. Strategies for preventing CIN [3]

Strategies that do not work
Mannitol
Loop diuretics
Dopamine
Atrial natriuretic peptide
Fenoldopam
Hemodialysis
Strategies that may work
Calcium channel blockers
Theophylline
Iso-osmolar contrast media
Hemofiltration
N-acetylcysteine
Sodium bicarbonate
Ascorbic acid
Prostaglandins
Currently recommended strategies
Noniodinated contrast
Avoid NSAIDs
Provide time between contrast doses
Minimize contrast volumes
Parenteral hydration
Use iso- or low osmolar media

dehydration and minimizing contrast osmolar load (type and volume) are currently the mainstays of CIN and CSA-AKI prevention. In the future, the defined temporal pattern of these two conditions presents an opportunity for investigators to test novel ideas and compounds that may later be extended to other AKI syndromes with less well-defined time courses.

References

1 Marenzi G et al: Contrast-induced nephropathy in patients undergoing primary angioplasty for acute myocardial infarction. J Am Coll Cardiol 2004;44:1780–1785.

2 Bentley ML, Corwin H L, Dasta J: Drug-induced acute kidney injury in the critically ill adult: recognition and prevention strategies. Crit Care Med 2010;38:S169–S174.

3 Rudnick MR, Kesselheim A, Goldfarb S: Contrast-induced nephropathy: how it develops, how to prevent it. Cleve Clin J Med 2006;73:75–80, 83–87.

4 Shaw A, Swaminathan M, Stafford-Smith M: Cardiac surgery-associated acute kidney injury: putting together the pieces of the puzzle. Nephron Physiol 2008;109:55–60.

5 Cruz DN, Bagshaw, SM, Ronco C, Ricci Z: Acute kidney injury: classification and staging; in Ronco C, Costanzo MR, Bellomo R, Maisel AS (eds): Fluid Overload: Diagnosis and Management. Contrib Nephrol. Basel, Karger, 2010, vol 164, pp 24–32.
6 Endre ZH, Westhuyzen J: Early detection of acute kidney injury: emerging new biomarkers. Nephrology (Carlton) 2008;13:91–98.
7 Bedi HS, Singh RP, Sengar B, Sachdeva G: Contrast-induced nephropathy and cardiac surgery. Indian Heart J 2004;56:262–263.
8 Heyman S N, Reichman J, Brezis M: Pathophysiology of radiocontrast nephropathy: a role for medullary hypoxia. Invest Radiol 1999;34:685–691.
9 Heyman SN, Rosen S, Brezis M: Radiocontrast nephropathy: a paradigm for the synergism between toxic and hypoxic insults in the kidney. Exp Nephrol 1994;2:153–157.
10 Heyman SN, et al: Radiocontrast agents induce endothelin release in vivo and in vitro. J Am Soc Nephrol 1992;3:58–65.
11 Rosenberger P, et al: The influence of epiaortic ultrasonography on intraoperative surgical management in 6051 cardiac surgical patients. Ann Thorac Surg 2008;85:548–553.
12 Haase M, et al. Sodium bicarbonate to prevent increases in serum creatinine after cardiac surgery: a pilot double-blind, randomized controlled trial. Crit Care Med 2009;37:39–47.

Andrew Shaw MB, FRCA, FCCM
Ass. Prof., Department of Anesthesiology
Duke University Medical Center
Durham, NC 27516 (USA)
Tel. +1 919 286 6938, E-Mail andrew.shaw@duke.edu

Kellum JA, Ronco C, Vincent J-L (eds): Controversies in Acute Kidney Injury.
Contrib Nephrol. Basel, Karger, 2011, vol 174, pp 163–172

Acute Kidney Injury in Children: Prevention, Treatment and Rehabilitation

Stuart L. Goldstein

University of Cincinnati College of Medicine, Director, Center for Acute Care Nephrology, Cincinnati Children's Hospital Medical Center, Cincinnati, Ohio, USA

Abstract

Pediatric acute kidney injury (AKI) epidemiology has shifted from primary kidney disease to secondary to another organ system illness or its treatment with nephrotoxic medications. Similar to adult patients, critically ill children with AKI with multiorgan failure exhibit high mortality rates, yet conducting interventional trials to prevent, treat or mitigate the effects of AKI in children have been hampered by relatively low event rates and the reliance on serum creatinine as the biomarker of AKI. However, recent advancements in standardizing the AKI definition via the pediatric modified RIFLE criteria, multicenter collaboration via the Prospective Pediatric CRRT Registry Group and multiple validation studies of novel AKI biomarkers in children have provided the essential components to evaluate preventive and therapeutic strategies to attack pediatric AKI as a disease state. The scope of this article is to review the advancements in the study of pediatric AKI over the past decade and offer a compelling and bright view of what is on the horizon for the prevention, treatment and rehabilitation of AKI in kids.

The landscape of pediatric acute kidney injury (AKI) has undergone a dramatic shift in the last 10 years. A reassessment of pediatric AKI epidemiology, standardization of the pediatric AKI definition, collaborative multicenter continuous renal replacement therapy (CRRT) research, assessment of CKD development in pediatric AKI survivors, and an intense effort to discover and validate novel urinary AKI biomarkers comprise major advancements in the field. Moreover, they serve as the necessary foundation to begin interventional trials to prevent or mitigate the effects of pediatric AKI, with an infrastructure to follow children longitudinally after an AKI episode. The purpose of this article is to review the

Table 1. Shift in pediatric AKI epidemiology

Author	Year	Time span	Cohort	AKI cause
Williams et al. [4]	2002	1978–1998	all-hospital	1978–1988: HUS (38%), oncology 8% 1988–1998: HUS (22%), oncology (17%)
Hui-Stickle et al. [3]	2005	1999–2001	all-hospital	ischemic (21%) nephrotoxins (16%) primary renal (7%)
Akcan-Arikan et al. [12]	2007	2005–2006	PICU	pneumonia (33%) SIRS/sepsis (27%) cardiogenic (10%)

PICU = Pediatric intensive care unit; HUS = hemolytic uremic syndrome; SIRS = systemic inflammatory reaction syndrome.

relevant pediatric AKI literature with a focus on the fast approaching horizon that holds promise for making a positive clinical impact for the child with, or at risk for, AKI.

Pediatric Acute Kidney Injury Epidemiology – The Definition Has an Impact on Outcome

The study of pediatric AKI epidemiology has intensified over recent years, likely as a result of more widespread provision of acute renal replacement therapy (RRT) modalities to critically ill children [1]. It appears that hospital and pediatric intensive care unit-acquired pediatric AKI rates have increased more than 9-fold from the 1980s through 2004 due to increasing use of more invasive management and higher illness severity of critically ill children [2]. Recent data [2–4] reveal a dramatic shift in the epidemiology of pediatric AKI (table 1), with the most common causes being renal ischemia, nephrotoxin use and sepsis. In addition, patients receiving prolonged courses of nephrotoxins are not being screened routinely for AKI development [5, 6]. Thus, pediatric AKI develops more often in hospitalized children as a result of a systemic illness or its treatment, and may be underreported due to low screening rates.

Another limitation of the past pediatric AKI literature was the lack of a standardized AKI definition. The incidence of the most severe forms of pediatric AKI, defined by dialysis requirement, ranges from 1 to 2% of all critically ill children [2, 7]. The incidence of AKI in children undergoing cardiopulmonary bypass is 10–50%, depending on the definition used [8–10]. Even small increases in serum creatinine are now recognized to contribute to poor pediatric patient outcomes. Children with acute decompensated heart failure and with a 0.3-mg/dl or greater serum creatinine rise demonstrate a 7-fold increased mortality risk [11].

In 2004, a standardized AKI consensus definition was proposed by the Acute Dialysis Quality Initiative: the RIFLE criteria, which were modified, applied and validated in pediatric patients and renamed the pediatric RIFLE (pRIFLE) criteria. pRIFLE stratifies pediatric AKI from mild (RIFLE-R, 'risk') to severe (RIFLE-F, 'failure') based on changes in serum creatinine or estimated creatinine clearance and urine output. The first study which defined AKI using the pRIFLE criteria found that AKI occurred in 82% of critically ill children admitted to a pediatric intensive care unit who received invasive mechanical ventilation and at least one vasoactive medication [12]. Worsening pediatric AKI defined by pRIFLE criteria was an independent risk factor for mortality and increased hospital length of stay. Subsequent reports confirmed the utility of the RIFLE and pRIFLE criteria to stratify pediatric AKI severity, highlighting the potential association between pediatric AKI severity and patient outcome [13–16]. After cardiopulmonary bypass, children demonstrate a stepwise increase in mortality and left ventricular dysfunction associated with increased AKI severity by pRIFLE [17, 18].

These studies validate the need for standardized AKI definitions for children as well as the need to focus on earlier detection of AKI before a patient requires RRT. Further prospective epidemiologic studies with detailed descriptions of the particular population studied will be crucial to understanding the true incidence of mild-to-severe AKI in a wide range of geographic and diagnostic patient populations. A recent publication proposed the concept of a 'renal angina equivalent' to prompt investigation into the presence and causes of AKI, much as chest pain and associated signs and symptoms prompt evaluation for acute coronary syndrome and myocardial infarction [19]. The renal angina concept comprises a set of testable hypotheses to risk-stratify patients, based on underlying systemic disease and illness severity, for interventional studies. An accurate identification of patients at risk for pediatric AKI, coupled with the multidimensional pRIFLE classification system, provides an essential foundation for preventive and therapeutic studies to attack pediatric AKI.

Multicenter Collaboration in Pediatric Acute Kidney Injury – Invention Born of Necessity

Care for the critically ill patient with sepsis and AKI is complicated by the need to manage multiorgan system failure, often requiring complex supportive measures of fluid resuscitation, vasoactive medication administration and decisions as to the timing of RRT. Clinical research in adults with sepsis and acute respiratory distress syndrome has also primarily focused on the benefits of early and aggressive goal-directed fluid resuscitation to restore end-organ provision, with some recent attention paid to conservative late fluid management strategies to limit fluid administration [20]. However, pediatric studies have examined the concept of fluid accumulation in the critically ill child with AKI. Children with AKI provide an informative population for study as their care is not usually complicated by comorbidities found in adults, such as atherosclerotic heart disease, diabetes or chronic obstructive pulmonary disease.

Can Fluid Be a Toxic Medication?

In the setting of AKI, physicians are very cognizant to limit the dose of potentially harmful electrolytes (potassium, phosphorus) provided in exogenous fluids, but the concept of a fluid volume dose has been limited for the most part to an acute dose to treat hypotension (e.g. 10 ml/kg of normal saline). Yet, the concept of a deleterious degree of positive fluid accumulation, or fluid overdose, has received no systemic evaluation. For example, neither of the two most recent, comprehensive, and important randomized and controlled trials comparing small solute dose of RRT (the ATN and RENAL trials) reported a positive fluid balance in their patient cohorts at the time of RRT initiation. Given that these patients had oligoanuric AKI and disordered fluid homeostasis is a primary indication to initiate RRT, our collective ignorance regarding the fluid balance status in patients with AKI is perplexing.

Why has cumulative fluid balance received such short shrift? Clinicians and AKI investigators may have assumed patients are getting the amount of fluid they need (and maybe too little, but rarely too much), and since it is usually of a relatively isotonic composition (e.g. normal saline or Ringer's lactate) and can be removed by RRT, they may have assumed that fluid cannot really be overdosed. However, lessons from the pediatric AKI literature challenge these assumptions.

The Innovative Lessons from the Pediatric Intensive Care Unit

The lessons from pediatric nephrologists and intensivists emanate from two practice perspectives ingrained into pediatricians: (1) disease prevention and (2) medication dosing based on patient size. These perspectives are neither unique to pediatrics nor absent in internal medicine, but they are more common in our training and everyday practice. In pediatric AKI and RRT, the concept of relative fluid accumulation (percent fluid overload) based on intensive care unit admission weight and timing of renal replacement based on percent fluid overload, not blood urea nitrogen concentration, has driven extensive pediatric research in the past decade.

Critically ill children often require aggressive fluid and inotropic support to maintain adequate perfusion. Substantial single-center and multicenter pediatric studies over the past decade have demonstrated that increasing degrees of relative fluid accumulation, or percent fluid overload, at the time of RRT initiation in children with AKI is independently associated with mortality [21–26]. Percent fluid overload (%FO) is calculated by totaling fluid volumes from ICU admission to RRT initiation using the following equation:

%FO = [(Fluid Input (l) – Fluid Output (l))/Patient intensive care unit admission weight (kg)].

In all of these studies, estimated glomerular filtration rate (GFR), patient age and size, urine output, diuretic use, and severity of illness did not differ between survivors and nonsurvivors. Analysis of different percent thresholds from these studies suggest mortality increases from 40 to 60% in children with more than 10–20% fluid overload at RRT initiation, independent of patient severity of illness. One could still possibly argue that the patients actually 'needed' the fluids they received. However, in a published multicenter study from the Prospective Pediatric CRRT Registry Group, the mean central venous pressure for survivors was 16.5 ± 6.1 mmH_20 versus 21.2 ± 6.6 mmH_2O for nonsurvivors [24]. Current recommendations for early goal-directed fluid resuscitation advocate fluid administration until a central venous pressure of 8. Since the mean central venous pressure was 2- to 3-fold above target recommendations, it is difficult to support the notion that patients received only the fluid volumes they needed and not an excess amount of fluid.

The Prospective Pediatric Continuous Renal Replacement Therapy Registry Group recently conducted an analysis of its entire 340-patient cohort using a tripartite classification for percent fluid overload at CRRT initiation: <10% FO, 10–20% FO and >20% FO [26]. The findings of this analysis confirmed an 8.5 OR of death for patients in the higher fluid overload group, irrespective of severity of illness. As important, the Prospective Pediatric Continuous Renal Replacement Therapy Registry Group collaboration has been essential for assessing multiple potential associations between clinical factors and patient outcome, as well as providing guidance for optimal vascular and anticoagulation management in children receiving CRRT. The collaboration was essential since even busy pediatric programs may only care for 30–40 patients requiring CRRT annually. The Prospective Pediatric CRRT Registry Group provides a model for multicenter network participation in pediatric acute kidney disease, and in the future, will not be solely focused on CRRT.

Acute Kidney Injury Biomarkers – It's Time to Use Them to Direct Care and Studies in Children Can Lead the Way

Even though the data from pediatric studies showing increased risk of morbidity and mortality in children with small rises in the serum of serum creatinine and/or moderate fluid overload are compelling, most clinicians are not willing to initiate RRT or fluid conservative strategies in patients with only a 0.3-mg/

dl serum creatinine rise or 10% positive fluid accumulation. There is always the hope that the patient can recover from their acute kidney disease without the need for dialysis. Many physicians do not consider initiating therapy even at a doubling of serum creatinine, which begs the question 'What other organ do we allow to lose more than 50% of its function before we change our management?' However, if clinicians had a reliable marker to predict AKI development, severity and/or duration, they may become more comfortable starting 'early'.

Much has been made of urinary AKI biomarkers to serve as the 'troponin I' to diagnose subclinical AKI early and be used in therapeutic trials to prevent or mitigate the effects of AKI. It is not the purpose of this article to review the plethora of pediatric AKI biomarker data, yet children provide an especially informative group for biomarker study and have led the way in biomarker validation studies after cardiopulmonary bypass, hemolytic uremic syndrome (HUS) and critical illness [27, 28]. Since, as noted before, children do not generally possess chronic confounding comorbidities, they can serve as the logical first translational step from bench to bedside in evaluating biomarker-directed therapies to prevent AKI. Only children risk-stratified by renal angina criteria, with elevated early biomarkers or markers of prolonged AKI to further stratify to higher AKI development or severity risk with early injury, should be enrolled in randomized preventive and therapeutic studies.

Recovery from Acute Kidney Injury Is Not What It Used to Be

The long-term sequelae of clinical pediatric AKI have been the subject of only recent investigation since, as noted above, no systematic assessment of pediatric survivors has been undertaken. Hui-Stickle et al. [3] demonstrated that 34% of 176 children had either reduced kidney function or were dialysis-dependent upon discharge from a tertiary center after a pediatric AKI. Askenazi et al. [29] followed this cohort for 3–5 years and found patient survival to be 56.8%, with the majority of mortality occurring within 2 years of the pediatric AKI episode. Sixteen children progressed to end-stage renal disease, with the majority (91%) having an underlying renal or urological disease. Thus, in the relative short term, progression to RIFLE-E in children with pediatric AKI may represent an acute or chronic kidney disease paradigm, although we do not have data to suggest that AKI accelerated the progression to end-stage renal disease in these patients. Seventeen of 29 patients studied in follow-up demonstrated evidence of CKD, manifesting as hyperfiltration, reduced kidney function, hypertension or microalbuminuria. The majority of these children with CKD had more than one CKD sign or symptom, yet only 6 were being followed by a pediatric nephrologist. This early small-scale study should prompt the pediatric nephrology community to perform systematic longitudinal evaluations for pediatric CKD in children who survive a pediatric AKI episode.

Garg et al. [30] performed an extensive meta-analysis of the long-term consequences of HUS. The authors included 49 studies comprising 3,476 patients for review. The primary endpoints were determining the incidence of, and risk factors for, the development of death, end-stage renal disease or CKD (defined as an estimated GFR <80 ml/min/1.73 m^2). Secondary renal endpoints included proteinuria and hypertension. The pooled estimates for CKD development were 8% (95% CI: 5–11) for GFR 60–80 ml/min/1.73 m^2, 6% (95% CI: 3–8) for GFR 30–59 ml/min/1.73 m^2 and 1.8% (95% CI: 0.8–3) for GFR 5–29 ml/min/1.73 m^2. In addition, the pooled rates for proteinuria and hypertension were 15% (95% CI: 10–20) and 10% (95% CI: 8–12), respectively. The combined renal sequelae rates showed 25% (95% CI: 20–30) of HUS survivors are at risk for long-term kidney damage. An important finding of this study was the observation of the development of long-term kidney sequelae in 8–61% of patients who had return of normal GFR and/or loss of proteinuria after the acute HUS episode. Thus, long-term follow-up for at-risk patients is warranted after an acute HUS episode.

What Is on the Horizon for Pediatric Acute Kidney Injury – Research Issues Looking Forward

The epidemiological standardization of pediatric AKI using pRIFLE and the renal angina concept, demonstration of effective multicenter collaboration via the Prospective Pediatric CRRT Registry Group, suggestion of an AKI-CKD progression and validation of novel biomarker precision to detect AKI development and severity early have been fundamental and essential developments to set the stage for evaluation of preventive and therapeutic strategies to prevent, treat and/or mitigate pediatric AKI. Once point-of-care AKI biomarkers are available to the clinical researcher, they can be used to stratify patients who are at a high risk of developing clinical AKI (fig. 1). The general concept for trial design is as follows:

1 Identify patients at high-risk for AKI based on the renal angina hazard tranches
2 Screen these patients for early subclinical AKI, with point-of-care novel biomarker assessments
3 Enroll *only* patients with elevated novel biomarker levels into:
   a. Trials to assess the efficacy of compounds directed at preventing clinical AKI
   b. Trials that direct clinical decision-making with respect to:
      i. Aggressive fluid resuscitation endpoints
      ii. Conservative late fluid management
      iii. Initiation of RRT
4 Follow all at-risk patients longitudinally to assess CKD development in:

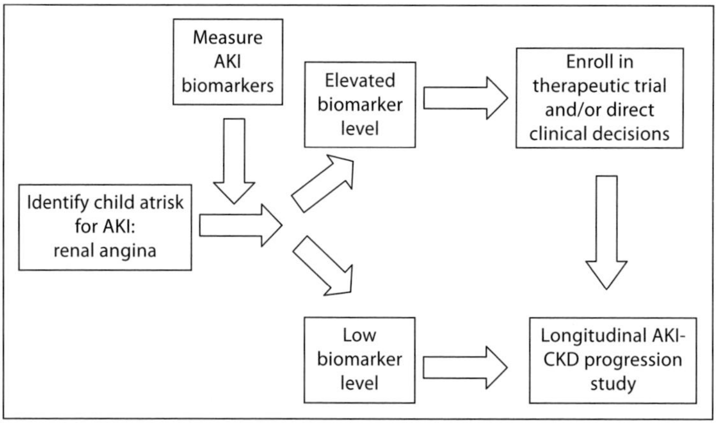

Fig. 1. Pediatric AKI research design paradigm: What's on the horizon?

   a. Patients with vs. without elevated biomarker levels
   b. Patients with vs. without AKI by pRIFLE (stratified based on maximum pRIFLE)
   c. The Translational Research Investigating Biomarker Endpoints for Acute Kidney Injury (TRIBE-AKI, www.yale.edu/tribeaki) Study consortium has formed an ancillary study group, the Assessment, Serial Evaluation, and Subsequent Sequelae in Acute Kidney Injury (ASSESS-AKI) Consortium, to examine long-term kidney outcomes in children with cardiopulmonary bypass associated AKI

5 Consider enrollment of children >13 years of age in all AKI trials since they represent a cohort without confounding clinical factors.

References

1 Warady BA, Bunchman T: Dialysis therapy for children with acute renal failure: survey results. Pediatr Nephrol 2000;15:11–13.
2 Vachvanichsanong P, Dissaneewate P, Lim A, McNeil E: Childhood acute renal failure: 22-year experience in a university hospital in southern Thailand. Pediatrics 2006;118:e786–e791.
3 Hui-Stickle S, Brewer ED, Goldstein SL: Pediatric ARF epidemiology at a tertiary care center from 1999 to 2001. Am J Kidney Dis 2005;45:96–101.
4 Williams DM, Sreedhar SS, Mickell JJ, Chan JC: Acute kidney failure: a pediatric experience over 20 years. Arch Pediatr Adolesc Med 2002;156:893–900.
5 Zappitelli M, Moffett BS, Hyder A, Goldstein SL: Acute kidney injury in non-critically ill children treated with aminoglycoside antibiotics in a tertiary healthcare centre: a retrospective cohort study. Nephrol Dial Transplant 2011;26:144–150.
6 Moffett BS, Goldstein SL: Pediatric acute kidney injury is associated with increasing nephrotoxic medication exposure. Clin J Am Soc Nephrol 2011;6:856–863.

7 Bailey D, Phan V, Litalien C, et al: Risk factors of acute renal failure in critically ill children: a prospective descriptive epidemiological study. Pediatr Crit Care Med 2007;8:29–35.

8 Mishra J, Dent C, Tarabishi R, et al: Neutrophil gelatinase-associated lipocalin (NGAL) as a biomarker for acute renal injury after cardiac surgery. Lancet 2005;365:1231–1238.

9 Dent CL, Ma Q, Dastrala S, et al: Plasma neutrophil gelatinase-associated lipocalin predicts acute kidney injury, morbidity and mortality after pediatric cardiac surgery: a prospective uncontrolled cohort study. Crit Care 2007;11:R127.

10 Bennett M, Dent CL, Ma Q, et al: Urine NGAL predicts severity of acute kidney injury after cardiac surgery: a prospective study. Clin J Am Soc Nephrol 2008;3:665–673.

11 Price JF, Mott AR, Dickerson HA, et al: Worsening renal function in children hospitalized with decompensated heart failure: evidence for a pediatric cardiorenal syndrome? Pediatr Crit Care Med 2008;9:279–284.

12 Akcan-Arikan A, Zappitelli M, Loftis LL, Washburn KK, Jefferson LS, Goldstein SL: Modified RIFLE criteria in critically ill children with acute kidney injury. Kidney Int 2007;71:1028–1035.

13 Plotz FB, Bouma AB, van Wijk JA, Kneyber MC, Bokenkamp A: Pediatric acute kidney injury in the ICU: an independent evaluation of pRIFLE criteria. Intensive Care Med 2008;34:1713–1717.

14 Palmieri T, Lavrentieva A, Greenhalgh D: An assessment of acute kidney injury with modified RIFLE criteria in pediatric patients with severe burns. Intensive Care Med 2009;35:2125–2129.

15 Zappitelli M, Parikh CR, Akcan-Arikan A, Washburn KK, Moffett BS, Goldstein SL: Ascertainment and epidemiology of acute kidney injury varies with definition interpretation. Clin J Am Soc Nephrol 2008;3:948–954.

16 Schneider J, Khemani R, Grushkin C, Bart R: Serum creatinine as stratified in the RIFLE score for acute kidney injury is associated with mortality and length of stay for children in the pediatric intensive care unit. Crit Care Med 2010;38:933–999.

17 Zappitelli M, Bernier PL, Saczkowski RS, et al: A small post-operative rise in serum creatinine predicts acute kidney injury in children undergoing cardiac surgery. Kidney Int 2009;76:885–892.

18 Blinder JJ, Goldstein SL, Lei VV, et al: Acute kidney injury occurs frequently in infants undergoing congenital heart surgery. J Thorac Cardiovasc Surg, submitted.

19 Goldstein SL, Chawla LS: Renal angina. Clin J Am Soc Nephrol 2010;5:943–949.

20 Murphy CV, Schramm GE, Doherty JA, et al: The importance of fluid management in acute lung injury secondary to septic shock. Chest 2009;136:102–109.

21 Foland JA, Fortenberry JD, Warshaw BL, et al: Fluid overload before continuous hemofiltration and survival in critically ill children: a retrospective analysis. Crit Care Med 2004;32:1771–1776.

22 Gillespie RS, Seidel K, Symons JM: Effect of fluid overload and dose of replacement fluid on survival in hemofiltration. Pediatr Nephrol 2004;19:1394–1399.

23 Goldstein SL, Currier H, Graf C, Cosio CC, Brewer ED, Sachdeva R: Outcome in children receiving continuous venovenous hemofiltration. Pediatrics 2001;107:1309–1312.

24 Goldstein SL, Somers MJ, Baum MA, et al: Pediatric patients with multi-organ dysfunction syndrome receiving continuous renal replacement therapy. Kidney Int 2005;67:653–658.

25 Hayes LW, Oster RA, Tofil NM, Tolwani AJ: Outcomes of critically ill children requiring continuous renal replacement therapy. J Crit Care 2009;24:394–400.

26 Sutherland SM, Zappitelli M, Alexander SR, et al: Fluid overload and mortality in children receiving continuous renal replacement therapy: the prospective pediatric continuous renal replacement therapy registry. Am J Kidney Dis 2010;55:316–325.

27 Zappitelli M, Washburn KK, Arikan AA, et al: Urine neutrophil gelatinase-associated lipocalin is an early marker of acute kidney injury in critically ill children: a prospective cohort study. Crit Care 2007;11:R84.
28 Washburn KK, Zappitelli M, Arikan AA, et al: Urinary interleukin-18 is an acute kidney injury biomarker in critically ill children. Nephrol Dial Transplant 2008;23:566–572.
29 Askenazi DJ, Feig DI, Graham NM, Hui-Stickle S, Goldstein SL: 3–5 year longitudinal follow-up of pediatric patients after acute renal failure. Kidney Int 2006;69:184–189.
30 Garg AX, Suri RS, Barrowman N, et al: Long-term renal prognosis of diarrhea-associated hemolytic uremic syndrome: a systematic review, meta-analysis, and meta-regression. JAMA 2003;290:1360–1370.

Stuart L. Goldstein, MD, Professor of Pediatrics
Cincinnati Children's Hospital Medical Center
Center for Acute Care Nephrology
3333 Burnet Avenue, Cincinnati, OH 45229 (USA)
Tel. +1 513 803 3295, E-Mail stuart.goldstein@cchmc.org

Kellum JA, Ronco C, Vincent J-L (eds): Controversies in Acute Kidney Injury.
Contrib Nephrol. Basel, Karger, 2011, vol 174, pp 173–181

Delayed Kidney Allograft Function – What Does It Tell Us about Acute Kidney Injury?

Bassam G. Abu Jawdeh · Hamid Rabb

Division of Nephrology, Johns Hopkins University School of Medicine, Baltimore, Md., USA

Abstract

Ischemia-reperfusion (IR) injury plays a significant role in the pathogenesis of both delayed graft function (DGF) in allografts and hemodynamic mediated acute kidney injury (AKI) of native kidneys. IR injury involves the interplay of several factors, including both the innate and adaptive immune systems. Though the final effector mechanisms causing injury are likely similar in both allograft and native kidneys, DGF is confounded by various donor, recipient and graft-handling factors. DGF is becoming increasingly important because of the growing number of patients awaiting kidney transplant. Identifying mediators, new diagnostics and therapeutics for DGF is important to provide more kidneys for transplant and improve outcomes. Furthermore, given that DGF occurs in a tightly controlled scenario, studying DGF is useful for early studies of proof of principle for biomarkers and therapeutics prior to launching into less well-controlled studies in large numbers of native kidney AKI patients.

Delayed graft function (DGF) is the term used to describe the inability of the kidney allograft to function post-transplantation, requiring dialysis. Although many factors may be involved in the pathogenesis of DGF, the term is commonly used interchangeably with ischemia-reperfusion (IR) injury. DGF is considered to be from IR injury when other etiologies such as acute rejection, compromised renal perfusion and urine outflow obstruction are ruled out. The frequency of DGF can vary from 4 to 10% in living donor transplants and may reach 50% in deceased donor transplants [1]. This high incidence is partly explained by the ongoing increased utilization of expanded criteria and post-cardiac death allografts. DGF is associated with an increased risk of acute rejection and long-term graft loss [1, 2], hence the increasing burden of this entity. In this review we will primarily discuss IR injury as the major cause of DGF, and briefly review the complex interplay between various donor and recipient factors which are unique to allografts when compared to acute kidney injury (AKI) of native kidneys.

Pathophysiology of Ischemia-Reperfusion Injury

The pathogenesis of IR injury is attributed to several processes triggered by an initial ischemic insult which is accentuated by reperfusion products. Our understanding of these processes has been mainly derived from animal studies, though some have been confirmed by observations in human transplants. Ischemia results in deprivation of oxygen and nutrients with concomitant accumulation of metabolic waste products. The absence of oxygen leads to ATP depletion which culminates in Na/K-ATPase inactivation and an increase in anaerobic metabolism leading to accumulation of lactic acid [1]. This in turn leads to reduced cytosolic pH and activation of lysosomal enzymes which catalyze the production of oxygen free radicals and nitric oxide. Under ischemic conditions, nitric oxide reacts with superoxide anion to form peroxynitrite which induces direct renal injury and is thus deleterious to the kidney. Subsequent reperfusion leads to reoxygenation and ATP production which is complicated by membrane lipoperoxidation thus inducing further cell death. Moreover, damaged endothelium is associated with excess production of thromboxane and endothelin and reduction in prostacyclin manifesting in vasoconstriction and further ischemia [3–5].

Both the innate and the adaptive immune systems play an active role in IR injury pathogenesis [6–8]. Reactive oxygen species generated by IR are accompanied by the generation of cytokines and expression of adhesion molecules on endothelial cells. This in turn recruits leukocytes leading to an amplification of the ongoing inflammation and thus cytotoxic injury [9]. Toll-like receptors and complement activation also play a role and can induce endothelial upregulation of adhesion molecules resulting in leukocyte aggregation and transmigration [10]. Renal tubular epithelial cells also play an active role in IR injury as they can engage with T cells, express adhesion molecules and produce proinflammatory cytokines [11–13]. This clear role of the immune system in IR injury could help explain the association between DGF and increased risk for rejection [14].

It was previously thought that the immune system was primarily involved in injury of the allograft during rejection, but not the pathogenesis of IR injury. Over the last several years, compelling data has demonstrated a key role for immunity in both DGF as well as native kidney IR injury. DGF and native kidney AKI share many common pathways, and an example of this is the common kidney injury biomarker findings in both disease states [15–17]. However, what confounds IR injury in transplant kidneys and makes it distinguishable from that in native kidneys is a combination of donor, graft handling and recipient factors that are discussed below.

Donor Factors

The shortage of organs for transplantation has led to the utilization of previously unaccepted kidneys to expand the donor pool and thus the number of transplant recipients. The United Network for Organ Sharing has defined expanded criteria donors as donors who are older than 60 years of age, or between 50 and 60 years of age and with two of the following characteristics: elevated serum creatinine levels above 1.5 mg/dl, history of systemic arterial hypertension and death from a cerebrovascular accident. The choice of these criteria was based on data that advocated an association between these donor factors and poor allograft outcomes including DGF [2, 18]. This is largely due to pre-existing donor kidney disease including glomerulosclerosis, interstitial fibrosis and tubular atrophy leading to a decrease in glomerular filtration rate and thus clearance function.

The shortage of organs has also led to the utilization of donation after cardiac death (DCD) kidneys. DCD kidneys are harvested from donors who die from cardiac arrest as opposed to brain dead donors where kidneys are harvested with a beating heart. DCD kidneys have significant but short periods of warm ischemia. The risk of DGF is 51% in DCD kidneys as opposed to 24% in standard-criteria donor kidney transplantation; however, long-term allograft and patient survival seem to be largely similar [18].

Brain death is a an independent predictor of graft failure which might explain, in addition to longer ischemia time, the inferior outcomes in recipients of cadaveric kidneys versus those of living donor kidneys [19]. At least two mechanisms appear to play a role. First is the increase in intracranial pressure associated with cerebrovascular accidents. This leads to the activation of the parasympathetic system which becomes opposed by an 'autonomic storm' (release of catecholamines) causing excessive vasoconstriction and relative hypoperfusion of the potential allograft. Subsequently, the parasympathetic system predominates, culminating in hypotension and further hemodynamic compromise of the kidney while still in the donor [20]. The second mechanism involves stress-related cytokine release and intrarenal inflammation. Upregulation of adhesion molecules and influx of leukocytes have been observed in kidneys of brain-dead rats irrespective of their hemodynamic status [21]. This observation has been corroborated in human kidneys from brain-dead donors with an increase in selectin adhesion molecules and leukocytes in the interstitium [22].

Graft Handling Factors

Storage conditions of donor kidneys play an important role in IR injury and DGF. Organ preservation has historically relied on two basic principles: (1) low-temperature preservation and (2) utilization of potassium-rich solutions [23]. Hypothermic storage at 0–4°C is used to decrease metabolism in the nonperfused organ. The duration of hypothermic storage is known as cold ischemia time (CIT), which is the time between the beginning of cold perfusion and the initiation of allograft vessel anastomosis in the recipient. As CIT increases the risk of DGF, the question 'What is a reasonable CIT in kidney transplantation?' arises. It appears that the incidence of DGF is influenced by CIT with an effect that seems to be continuous; however, efforts should focus on reducing CIT to less than 24 h when possible.

The high extracellular potassium concentration in preservation solutions is believed to maintain intracellular ionic content in the face of impaired function of ionic pumps during the preservation process, though this concept has been debated [24]. High potassium concentration induces cellular depolarization diminishing cellular ATP content. Moreover, voltage-dependent calcium channels are activated leading to an influx of calcium with secondary vasoconstriction leading to impaired allograft perfusion postanastomosis and reperfusion in the recipient [24]. This led to the development of low potassium solutions such as HTK (histidine, tryptophan and ketoglutarate) solution as opposed to the earlier developed high potassium UW (University of Wisconsin) solution. In a study comparing HTK to UW solutions, the HTK arm had a lower incidence of DGF when CIT was greater than 16 h. Graft and patient survival were not significantly different however [25].

The inhibition of sodium potassium ATPase during organ preservation is followed by accumulation of sodium intracellularly resulting in cell swelling. This mechanism was the basis for using preservation solutions with high concentrations of osmotically active substances to counteract this effect. Yet another deleterious consequence from ATP shortage is the inhibition of the calcium ATPase and activation of the sodium-calcium exchanger in reverse mode inducing intracellular calcium overload. This is associated with the opening of mitochondrial permeability transition pores, culminating in cellular apoptosis [23, 24]. The mitochondria contributes to DGF by also being a major source of oxygen free radicals, the primary effector molecules in IR injury as outlined above [24].

The membranes of the endothelial cells are associated with glycoproteins, glycolipids, glycosaminoglycans and proteoglycans comprising the glycocalyx. Glycocalyx limits the exposure of allogeneic molecules at the surface of endothelial cells, thus ameliorating immunologic injury and protecting against cell swelling [24]. During IR injury the integrity of the glycocalyx is altered, which predisposes the allograft to further injury and thus worsening DGF [24, 26]. Based on that, polymers such as polyethylene glycol could be superior to hydroxyethyl starch in preservation solutions as polyethylene glycol can restore glycocalyx function and contribute to immunocamouflage by interacting with the surface of donor renal cells and concealing the allogeneic sites from T cell receptors [27]. Although polyethylene glycol is cleared shortly after transplantation, its allogeneic modulatory effect is prolonged, which again underscores the importance of the immune system in DGF [28].

Machine perfusion is a technique that has been introduced in an attempt to allow longer preservation times. Kidneys are flushed and separated and then placed on a perfusion machine that pumps cold preservation solution continuously through the renal artery until time of transplantation. Two recent multicenter randomized controlled trials looking at kidneys from DCD donors revealed conflicting results [29, 30]. In the initial study, there was no difference in the incidence of DGF between kidneys assigned to hypothermic machine perfusion versus cold storage alone (58 vs. 56%, respectively) in the context of median CIT of 14 h [29]. In the subsequent trial, machine perfusion reduced the incidence of DGF from 69.5 to 53.7% (adjusted OR: 0.43; $p = 0.025$), and DGF was 4 days shorter in recipients of machine-perfused kidneys. Machine-perfused kidneys had a higher creatinine clearance up to 1 month after transplantation ($p = 0.027$). However, 1-year graft and patient survival were similar in both groups [30].

Postoperative and Recipient Factors

Postoperative hemodynamic compromise is one of the most common causes of ischemic injury in kidney allografts. Reduced renal blood flow augments the already existing damage from cold ischemia and it is therefore crucial to maintain adequate perfusion to the transplanted allograft. Prior hemodialysis is a risk factor for DGF as preemptive (prior to needing dialysis) transplantation has been associated with a lower rate of DGF compared with nonpreemptive transplantation for both deceased donor (8 vs. 26%; $p < 0.001$) and living donor transplants (2.6 vs. 6.1%; $p < 0.001$). The exact reason for this advantage of preemptive transplantation is not entirely clear; however, it could be explained by a set of confounders such as avoidance of cardiovascular disease that may have otherwise developed on maintenance hemodialysis and adequate residual function in native kidneys [31]. This could also explain why peritoneal dialysis patients, who usually maintain residual native kidney function, experience less DGF compared to those on hemodialysis [32]. Another important element is the use of calcineurin inhibitors immediately after transplantation. Calcineurin inhibitors have a narrow therapeutic range above which they can induce a dose-related thromboxane and endothelin-mediated vasoconstrictive effect, particularly on the afferent arterioles [33]. This leads to a prerenal AKI-like picture and DGF that is often confused with early rejection, and often warrants an allograft biopsy. Alternatively, immunosuppressive medications could positively modulate the immunologic injury caused by IR injury.

Rapamycin, an m-TOR inhibitor, was initially considered as an alternative to calcineurin inhibitors, hence minimizing their nephrotoxic effects; however, it has gradually become less used, particularly after the ELITE-Symphony trial which demonstrated superiority of low-dose tacrolimus (calcineurin inhibitor) for renal function and allograft survival. The incidence of DGF at 2 weeks was lower in the rapamycin group. However, the trial was not powered to examine DGF as a primary outcome [34]. The variants discussed above in addition to others such as donor age and body mass index have been incorporated in a logistic regression model that predicts DGF [2].

Conclusion

In comparing DGF to hemodynamic mediated AKI in native kidneys, we find that the most prominent end-effector mechanism in both cases is IR injury. Both entities are also associated with poorer long-term kidney outcomes. What makes DGF unique compared to native kidney AKI is the various donor, graft-handling and recipient factors discussed above.

The transplant community has developed protocols to approach each of these factors to best select a donor, recipient or organ for a transplant procedure. Therefore, although seeming like potential hurdles, they actually serve as a checklist to ensure the reproducibility of the transplant procedure in a safe and controlled setting. For example, given the increasing demand for kidney allografts, the procurement and utilization of cadaveric kidneys from donors with nonbeating hearts will continue, hence predisposing to IR injury and DGF. This can be offset, however, by prudent handling of organs, minimizing CIT and optimizing perioperative hemodynamics.

In native kidneys, however, AKI is frequently a result of a process that has started well before diagnosis, particularly in elderly patients with hemodynamic mediated AKI in a setting of intravascular volume depletion or infections. Furthermore, native kidneys usually require considerable injury to raise the serum creatinine before a diagnosis of AKI, which in turn delays intervention. The application of new biomarkers should improve this. Moreover, although patients undergoing kidney transplantation generally have comorbidities such as coronary artery disease, diabetes and hypertension, it is common practice to correct, stabilize or better control their chronic diseases prior to transplant surgery. The immune system might play a more prominent role in DGF as compared to native kidney AKI, but this is partially negated by introducing immunosuppressive medications at the time of transplantation. A rodent study has demonstrated that initiation of immunosuppression before transplantation can significantly attenuate IR injury [35]. Understanding and studying DGF is a useful approach to help elucidate novel mechanisms, diagnostics and therapeutics for transplant recipients as well as those with native kidney AKI.

References

1 Yarlagadda SG, Coca SG, Garg AX, Doshi M, Poggio E, Marcus RJ, Parikh CR: Marked variation in the definition and diagnosis of delayed graft function: a systematic review. Nephrol Dial Transplant 2008;23:2995–3003.

2 Irish WD, Ilsley JN, Schnitzler MA, Feng S, Brennan DC: A risk prediction model for delayed graft function in the current era of deceased donor renal transplantation. Am J Transplant 2010;10:2279–2286.

3 Wilhelm SM, Simonson MS, Robinson AV, Stowe NT, Schulak JA: Endothelin up-regulation and localization following renal ischemia and reperfusion. Kidney Int 1999;55:1011–1018.

4 Huang C, Huang C, Hestin D, Dent PC, Barclay P, Collis M, Johns EJ: The effect of endothelin antagonists on renal ischaemia-reperfusion injury and the development of acute renal failure in the rat. Nephrol Dial Transplant 2002;17:1578–1585.

5 Schilling M, Holzinger F, Friess H, Seiler C, Büchler MW: Pathogenesis of delayed kidney graft function: role of endothelin-1, thromboxane B2, and leukotriene B4. Transplant Proc 1996;28:304–305.

6 Jang HR, Rabb H: The innate immune response in ischemic acute kidney injury. Clin Immunol 2009;130:41–50.

7 Satpute SR, Park JM, Jang HR, Agreda P, Liu M, Gandolfo MT, Racusen L, Rabb H: The role for T cell repertoire/antigen-specific interactions in experimental kidney ischemia reperfusion injury. J Immunol 2009;183:984–992.

8 Chadban SJ, Wu H, Hughes J: Macrophages and kidney transplantation. Semin Nephrol 2010;30:278–289.

9 Bonventre JV, Weinberg JM: Recent advances in the pathophysiology of ischemic acute renal failure. J Am Soc Nephrol 2003;14:2199–2210.

10 Homeister JW, Lucchesi BR: Complement activation and inhibition in myocardial ischemia and reperfusion injury. Annu Rev Pharmacol Toxicol 1994;34:17–40.

11 Segerer S, Nelson PJ, Schlöndorff D: Chemokines, chemokine receptors, and renal disease: from basic science to pathophysiologic and therapeutic studies. J Am Soc Nephrol 2000;11:152–176.

12 Kapper S, Beck G, Riedel S, Prem K, Haak M, van der Woude FJ, Yard BA: Modulation of chemokine production and expression of adhesion molecules in renal tubular epithelial and endothelial cells by catecholamines. Transplantation 2002;74:253–260.

13 Li H, Nord EP: CD40 ligation stimulates MCP-1 and IL-8 production, TRAF6 recruitment, and MAPK activation in proximal tubule cells. Am J Physiol Renal Physiol 2002;282:F1020–F1033.

14 Howard RJ, Pfaff WW, Brunson ME, Ramos EL, Peterson JC, Croker BP, Scornik JC, Parris CJ, Fennell RS: Delayed graft function is associated with an increased incidence of occult rejection and results in poorer graft survival. Transplant Proc 1993;25:884.

15 Perry TE, Muehlschlegel JD, Liu KY, Fox AA, Collard CD, Shernan SK, Body SC, CABG Genomics Investigators: Plasma neutrophil gelatinase-associated lipocalin and acute postoperative kidney injury in adult cardiac surgical patients. Anesth Analg 2010;110:1541–1547.
16 Du Y, Zappitelli M, Mian A, Bennett M, Ma Q, Devarajan P, Mehta R, Goldstein SL: Urinary biomarkers to detect acute kidney injury in the pediatric emergency center. Pediatr Nephrol 2011;26:267–274.
17 Haase M, Haase-Fielitz A, Bellomo R, Mertens PR: Neutrophil gelatinase-associated lipocalin as a marker of acute renal disease. Curr Opin Hematol 2010, E-pub ahead of print.
18 Rao PS, Ojo A: The alphabet soup of kidney transplantation: SCD, DCD, ECD – fundamentals for the practicing nephrologist. Clin J Am Soc Nephrol 2009;4:1827–1831.
19 Roodnat JI, van Riemsdijk IC, Mulder PG, Doxiadis I, Claas FH, IJzermans JN, van Gelder T, Weimar W: The superior results of living-donor renal transplantation are not completely caused by selection or short cold ischemia time: a single-center, multivariate analysis. Transplantation 2003;75:2014–2018.
20 Pratschke J, Wilhelm MJ, Kusaka M, Basker M, Cooper DK, Hancock WW, Tilney NL: Brain death and its influence on donor organ quality and outcome after transplantation. Transplantation 1999;67:343–348.
21 van der Hoeven JA, Molema G, Ter Horst GJ, Freund RL, Wiersema J, van Schilfgaarde R, Leuvenink HG, Ploeg RJ: Relationship between duration of brain death and hemodynamic (in)stability on progressive dysfunction and increased immunologic activation of donor kidneys. Kidney Int 2003;64:1874–1882.
22 Nijboer WN, Schuurs TA, van der Hoeven JA, Leuvenink HG, van der Heide JJ, van Goor H, Ploeg RJ: Effects of brain death on stress and inflammatory response in the human donor kidney. Transplant Proc 2005;37:367–369.
23 Belzer FO, Southard JH: Principles of solid-organ preservation by cold storage. Transplantation 1988;45:673–676.
24 Hauet T, Eugene M: A new approach in organ preservation: potential role of new polymers. Kidney Int 2008;74:998–1003.
25 Agarwal A, Murdock P, Fridell JA: Comparison of histidine-tryptophan ketoglutarate solution and University of Wisconsin solution in prolonged cold preservation of kidney allografts. Transplantation 2006;81:480–482.
26 Rubio-Gayosso I, Platts SH, Duling BR: Reactive oxygen species mediate modification of glycocalyx during ischemia-reperfusion injury. Am J Physiol Heart Circ Physiol 2006;290:H2247–H2256.
27 Thuillier R, Renard C, Rogel-Gaillard C, Demars J, Milan D, Forestier L, Ouldmoulene A, Goujon JM, Badet L, Hauet T: Effect of polyethylene glycol-based preservation solutions on graft injury in experimental kidney transplantation. Br J Surg 2011;98:368–378.
28 Gallucci S, Matzinger P: Danger signals: SOS to the immune system. Curr Opin Immunol 2001;13:114–119.
29 Watson CJ, Wells AC, Roberts RJ, Akoh JA, Friend PJ, Akyol M, Calder FR, Allen JE, Jones MN, Collett D, Bradley JA: Cold machine perfusion versus static cold storage of kidneys donated after cardiac death: a UK multicenter randomized controlled trial. Am J Transplant 2010;10:1991–1999.
30 Jochmans I, Moers C, Smits JM, Leuvenink HG, Treckmann J, Paul A, Rahmel A, Squifflet JP, van Heurn E, Monbaliu D, Ploeg RJ, Pirenne J: Machine perfusion versus cold storage for the preservation of kidneys donated after cardiac death: a multicenter, randomized, controlled trial. Ann Surg 2010;252:756–764.
31 Kasiske BL, Snyder JJ, Matas AJ, Ellison MD, Gill JS, Kausz AT: Preemptive kidney transplantation: the advantage and the advantaged. J Am Soc Nephrol 2002;13:1358–1364.
32 Snyder JJ, Kasiske BL, Gilbertson DT, Collins AJ: A comparison of transplant outcomes in peritoneal and hemodialysis patients. Kidney Int 2002;62:1423–1430.

33 Olyaei AJ, de Mattos AM, Bennett WM: Nephrotoxicity of immunosuppressive drugs: new insight and preventive strategies. Curr Opin Crit Care 2001;7:384–389.
34 Ekberg H, Tedesco-Silva H, Demirbas A, Vítko S, Nashan B, Gürkan A, Margreiter R, Hugo C, Grinyó JM, Frei U, Vanrenterghem Y, Daloze P, Halloran PF, ELITE-Symphony Study: Reduced exposure to calcineurin inhibitors in renal transplantation. N Engl J Med 2007;357:2562–2575.
35 Cicora F, Lausada N, Vasquez DN, Cicora P, Guerrieri D, Gonzalez P, Zalazar G, Stringa P, Raimondi C: Protective effect of immunosuppressive treatment before orthotopic kidney autotransplantation. Transpl Immunol 2011;24:107–112.

Hamid Rabb, MD
Johns Hopkins Hospital
Ross 965, 720 Rutland Ave.
Baltimore, Maryland, 21205 (USA)
Tel. +1 410 502 1555, E-Mail hrabb1@jhmi.edu

Kellum JA, Ronco C, Vincent J-L (eds): Controversies in Acute Kidney Injury.
Contrib Nephrol. Basel, Karger, 2011, vol 174, pp 182–190

Acute Kidney Injury Leading to Chronic Kidney Disease and Long-Term Outcomes of Acute Kidney Injury: The Best Opportunity to Mitigate Acute Kidney Injury?

Lakhmir S. Chawla[a,b]

[a]Department of Anesthesiology and Critical Care Medicine, and [b]Division of Renal Diseases and Hypertension, Department of Medicine, George Washington University Medical Center, Washington DC, USA

Abstract

Acute kidney injury (AKI) has been shown to be associated with progression to chronic kidney disease (CKD). Multiple studies have shown that subsets of AKI survivors are at high risk for progression to advanced stage CKD and death. Risk factors associated with AKI survivors progressing to CKD have been identified and include advanced age, diabetes mellitus, decreased baseline glomerular filtration rate, severity of AKI and a low concentration of serum albumin. These risk factors can be utilized to identify those patients at highest risk for progression. Because progression to CKD in these AKI survivors typically occurs months after the initial AKI insult, a common injury pathway of CKD progression is probable, and therapeutic interventions that have been shown to retard CKD progression are likely to be effective in patients who survive AKI and then progress to CKD. AKI has many negative impacts across the spectrum of the disease. The 30-day mortality for patients with AKI is high, hence the preference to target AKI during the initiation phase. However, this phase is the most difficult point to treat AKI. The maintenance phase of AKI is longer in duration in comparison to the initiation phase, and thus the logistics are more amenable to study. However, the mainstay of treatment for the maintenance phase of AKI (renal replacement therapy) has been tested extensively and increasing the dose of renal replacement therapy has not been shown to improve outcome. Therefore, the recovery phase of AKI may represent the best opportunity to intervene in the negative outcomes of AKI.

Acute kidney injury (AKI) is becoming increasingly common in critically ill patients, and those patients with the most severe form of AKI requiring renal replacement therapy (RRT) have a mortality of 50–80% [1]. Over the past 10 years, there has been substantial progress in the field of AKI. In particular, consensus definitions, AKI biomarkers and the appropriate dosing of RRT have been established. Despite these advances, no drug or other therapeutic intervention currently exists for the treatment of AKI [2].

AKI is a clinical syndrome with a multitude of causes, which contributes to the complexities in designing therapies [3]. In addition, AKI has distinctive phases of injury (initiation, maintenance and recovery) [4]. While the consensus definition of AKI (RIFLE/AKIN) does not factor in the cause of AKI, it is likely that successful treatments for AKI will need to address the pathophysiologic mechanism of injury, particularly during the initiation phase [4].

AKI has been shown to be associated with progression to CKD [5–7]. Multiple studies assessing various cohorts of patients have shown that subsets of AKI survivors are at high risk for progression to advanced stage CKD [5–8]. In these studies, subsets of AKI survivors recovered renal function and then progressed to advanced stage CKD. Because the progression to CKD in these AKI survivors typically occurred months after the initial AKI insult, a common injury pathway of CKD progression is more probable, and therapeutic interventions that have been shown to forestall CKD progression are likely to be effective in patients who survive AKI and then progress to CKD.

Assessing the current landscape of AKI has shown that it has many negative impacts across the spectrum of the disease. Because the 30-day mortality for patients with AKI is 50–80%, there is an obvious desire to target AKI during the initiation phase [1]. However, this phase is the most difficult point to treat AKI. Despite the current progress in AKI biomarkers, real-time point-of-care tests for study intervention are not routinely available, except for a few select academic centers. The time window for intervention during the AKI initiation phase is short, and since these patients often cannot provide consent for themselves, the logistics of conducting these studies are daunting. These obstacles are further exacerbated by the fact that the understanding of the pathophysiologic mechanisms for the most common cause of AKI (sepsis-associated AKI, which contributes to half of all cases) is wanting [1, 9]. Moreover, the associated acute phase mortality of AKI is difficult to dissociate from the underlying disease.

The maintenance phase of AKI is longer in duration in comparison to the initiation phase, and thus the logistics are more amenable to study. However, the mainstay of treatment for the maintenance phase of AKI is RRT, which has been tested extensively. Two large well-powered prospective trials have shown that increasing the dose of RRT does not improve outcomes [10, 11]. Therefore, the

Table 1. Summary of studies indicating association of AKI with progression to CKD

Study	n	Key findings
Ali [26] 2007	562	AKI-F → 51% recover full renal function ACRF-F → 28% recover full renal function
Ishani [6] 2009	233,803	AKI → 13 times more likely to develop ESRD ACRF → 41 times more likely to develop ESRD
Hsu [28] 2009	1061	ACRF-F → 42–63% of hospital survivors progressed to ESRD
Lo [7] 2009	703	AKI-F survivors are 28 times more likely to develop advanced CKD
Amdur [5] 2009	5,404	20% of patients with a diagnosis of ATN progress to CKD4 within 20 months
Wald [8] 2009	3,769	AKI-F survivors are 3 times more likely to develop ESRD

AKI-F = Acute kidney injury RIFLE F; ACRF = acute chronic renal failure; ATN = acute tubular necrosis.

recovery phase of AKI may represent the best opportunity to intervene in the negative outcomes of AKI.

Acute Kidney Injury as a Cause of Chronic Kidney Disease

Prior to the era of CKD staging, it was generally accepted that patients who recovered from AKI had 'good' renal outcomes as assessed by a low incidence of end-stage renal disease (ESRD) [12]. These previous long-term studies were hampered by different definitions, variable follow-up times, and low overall numbers. More recently, from 2008 through 2010 (table 1), multiple studies assessing unique cohorts of patients demonstrated that patients who survive an episode of AKI have a significant risk for the development of advanced CKD (stage 4/5) [5–8].

Ishani et al. [6] assessed a random sample of Medicare beneficiaries and found that patients who had suffered an episode of AKI were more likely to develop ESRD than patients who did not have a history of AKI or CKD. In their analysis, the risk of ESRD was thirteen times higher than patients without a history of AKI or CKD. Because this study utilized diagnostic codes, it is difficult to determine the proportion of patients who suffered AKI and then progressed directly to ESRD as compared to patients who suffered AKI, recovered renal function and then progressed to ESRD [6]. Lo et al. [7], utilizing a database from the health insurer Kaiser Permanente, retrospectively studied inpatient admissions. Patients who suffered AKI (dialysis-requiring episode of AKI) were compared to patients who did not suffer an episode of AKI. Patients with a history of

ESRD or a pre-admission estimated glomerular filtration rate <45 ml/min/1.73 m^2 were excluded. In this study, the investigators found that an episode of dialysis requiring AKI was associated with a 28-fold increase of developing advanced CKD and a twofold increase in mortality. Because patients who developed ESRD within 30 days of discharge were excluded from the long-term follow-up of advanced CKD, these data are consistent with a severe form of AKI followed by renal recovery and then progression to advanced stage CKD [7]. Wald et al. [8] conducted a population-based cohort study of patients in Ontario, Canada with AKI patients who required in-hospital dialysis and survived free of dialysis for at least 30 days after discharge. These patients were matched with patients without AKI or dialysis during their index hospitalization. Patients with AKI requiring dialysis were over three times more likely to develop ESRD as compared to controlled matched patients.

We assessed a United States Department of Veterans Affairs database to ascertain long-term renal function in over 79,000 hospitalized patients [5]. Within this cohort, 63,491 patients were hospitalized for acute myocardial infarction or pneumonia without AKI and designated as controls. Another 5,404 hospitalized patients, which comprised the AKI group, had diagnostic codes indicating acute renal failure or acute tubular necrosis. Over the 5-year follow-up period, renal function deteriorated over time in all groups, but with significantly greater severity in those who had acute renal failure and acute tubular necrosis compared to controls. Patients with AKI, especially those with acute tubular necrosis, were more likely than controls to enter stage 4/5 CKD. We found that patients who had an episode of AKI were at high risk for the development of stage 4 disease and had a reduced survival time when compared to control patients [5].

In aggregate, these studies underscore the strong association between AKI and the future development of CKD.

Identification of Patients and Risk Factors for Chronic Kidney Disease Progression

The above studies in AKI survivors identified advanced age, presence of diabetes mellitus and decreased baseline estimated glomerular filtration rate as risk factors for the progression to advanced stage CKD. More recently, severity of AKI has been linked to CKD progression in survivors of AKI. Ishani et al. [13] assessed patients who underwent cardiac surgery and found that the magnitude of serum creatinine during the postoperative hospital course was directly linked to progression. This effect was seen in patients with de novo AKI as well as in patients with acute on chronic renal failure [13]. Similarly, we have recently shown that in addition to severity of AKI, a low serum albumin concentration is a strong predictor of poor long-term renal outcome [14]. The strong predictive value of serum albumin levels as a predictor of CKD progression is not surprising because low albumin levels have been associated with poor outcomes in a variety of diseases including ESRD, surgical illness and acute stroke [15–17]. Low levels of serum albumin can be due to either nutrition-related factors and/or high levels of inflammation [18]. Given that multiple studies have linked inflammation to AKI, it is likely that the predictive value of low concentrations of serum albumin to CKD progression is a composite measure of increased inflammation [19–21].

We assessed a cohort of patients with a spectrum of AKI from RIFLE stages R through F, and showed that the severity of AKI is associated with progression to advanced CKD (fig. 1). In particular, we showed that patients who require dialysis are at much higher risk for progression to CKD than patients with less severe AKI. The precise mechanism by which AKI causes CKD is unknown. Preclinical studies suggest several potential mechanisms for how AKI can cause CKD: acute endothelial injury leading to vascular dropout, nephron loss followed by glomerular hypertrophy or development of fibrosis after sustaining AKI [22, 23].

In order to help clinicians identify AKI survivors at high-risk for CKD progression, we developed three multivariable models. The most accessible model is based on sentinel clinical events (RIFLE stage, need for dialysis, etc.) [14]. This model performed well with 11% of the CKD variance explained (area under ROC curve = 0.77). We validated this equation in a separate large cohort of hospitalized veterans which resulted in $p < 0.001$ and c-statistics ≥0.81, signifying good reproducibility. Because these endpoints are accessible, clinicians could potentially use this equation to risk stratify AKI survivors who are at the highest risk for progression to CKD [14].

Interventions during the Recovery Phase of Acute Kidney Injury

Because a primary care provider sees most AKI patients following an AKI episode, these patients are accessible for intervention. At 30 days after AKI discharge, 74.5% have seen a primary physician compared to 11.9 and 29.5%, respectively, who see a nephrologist or cardiologist [24]. Surprisingly, only about one third of AKI patients who suffer AKI requiring dialysis see a nephrologist within 30 days of discharge. This increases to 48.6% within 1 year of discharge, which is likely a result of rising serum creatinine [24]. Overall, a very small fraction of AKI patients are currently followed by a nephrologist after an AKI discharge [24]. To put this into context, an assessment of the nephrology follow-up of AKI survivors as compared to the cardiology follow-up of myocardial infarction survivors reveals a stark difference (11.9 vs. 76%, respectively) [24, 25].

The public health impact of the long-term outcomes of AKI is significant. The population incidence of AKI is approximately 2,100 pmp [26]. Given

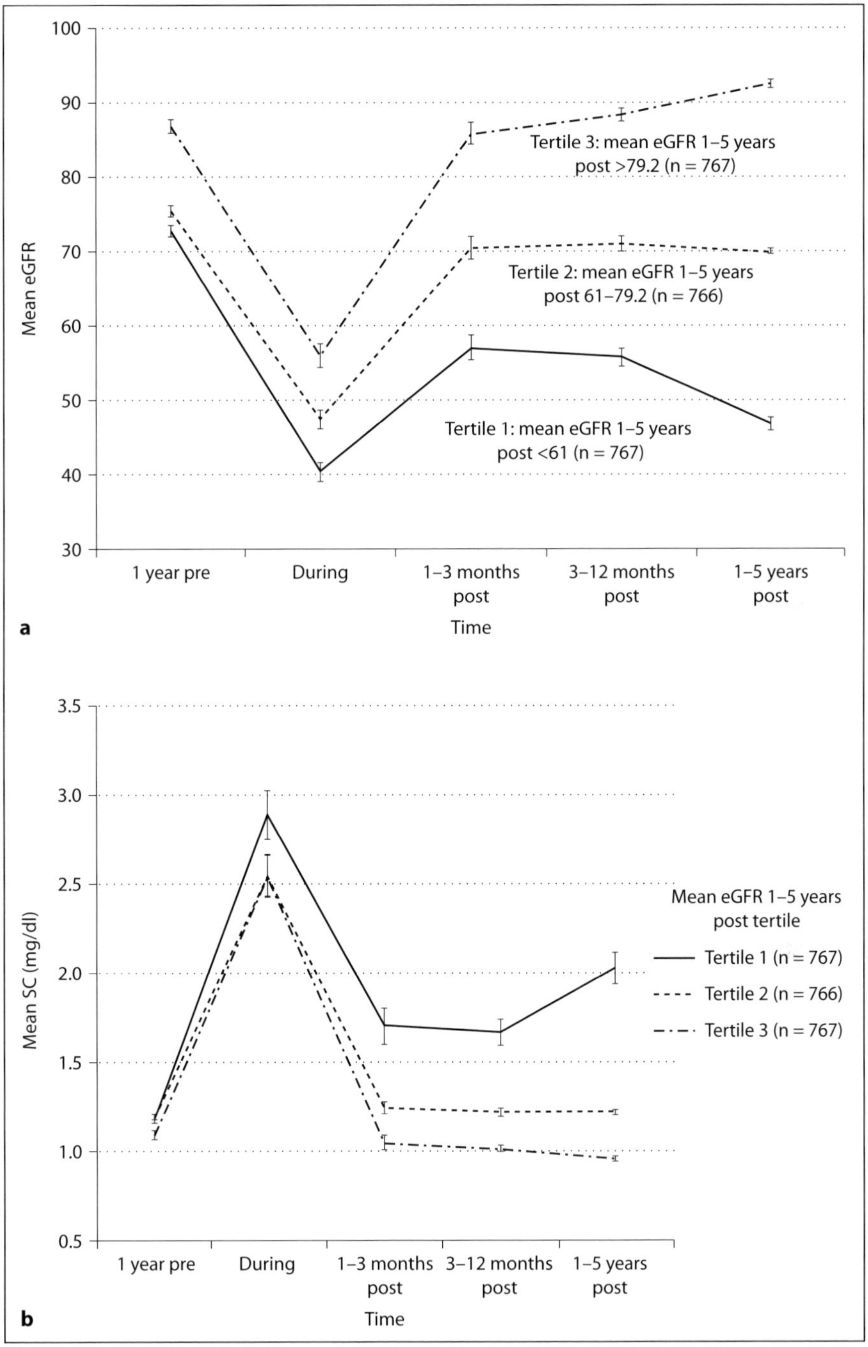

Fig. 1. Patients who were still alive ≥1 year after index hospitalization. **a** Time vs. mean estimated glomerular filtration rate (eGFR) in AKI survivors [14]. **b** Time vs. mean serum creatinine (SC) in AKI survivors [14].

the population of the developed world (USA, Canada, Western Europe and Australia) of approximately 1 billion, this means that there will be over 2 million cases of AKI this year, with an expected 1.5 million AKI survivors. Of these patients, approximately 15–20% will progress to advanced stage CKD within 24 months, resulting in approximately 300,000–400,000 cases of advanced CKD per year. Given the increasing incidence of AKI in the aging population [27], the numbers above are expected to increase.

In our view, the first interventional study to mitigate the progression of CKD in AKI survivors should include testing the standard approach that is currently used to treat CKD patients: blood pressure control, nephrotoxin avoidance and nutritional intervention. A potential study design would use a risk factor risk score to identify the patients at highest risk for CKD progression. Those patients would then be enrolled into a clinical trial during their index hospitalization of AKI. Patients would be randomized to one of three interventions: intervention 1, standard primary care follow-up at scheduled visits; intervention 2, standard primary care follow-up, with scheduled nephrology follow-up for blood pressure control, nutritional intervention for sodium and protein diet counseling, and nephrotoxin avoidance education; and intervention 3, interventions 1 and 2, plus renin-angiotensin-aldosterone system blockade. Other interventions that could be added or substituted in the above example include HMG Co-A reductase inhibitors, and antiproliferative agents.

Conclusions

Patients with AKI suffer a high short-term mortality rate and subsets of those patients who survive are at increased risk for developing CKD. Interventions targeted during the initiation phase of AKI are the most desirable. However, the inadequate understanding of the pathophysiologic mechanisms of this phase of AKI in combination with a paucity of candidate therapeutic agents underlies the complexities of bringing a successful intervention from the bench to the bedside.

While the recovery phase of AKI may be less appealing as a point to intervene because of the associated high short-tem mortality of AKI, the advantages of intervening in the recovery phase of AKI are numerous. The mechanism of CKD progression is more remote to the initial AKI. Patients at risk can be identified. The pathophysiology of CKD progression is well understood, and similarities between CKD progression and AKI progression to CKD are likely. The time to progression from AKI to recovery to advanced stage CKD occurs over months. Because of this longer timeline, incremental dose titration of potential therapeutic agents can be done more effectively, thereby potentially decreasing adverse events. Further, AKI survivors currently have no standard of care intervention, and modest interventions could have significant effects.

In short, intervention at the recovery stage of AKI possesses the important prerequisites for a successful intervention and this should be a focus for the AKI research community.

References

1 Uchino S, Kellum JA, Bellomo R, Doig GS, Morimatsu H, Morgera S, Schetz M, Tan I, Bouman C, Macedo E, Gibney N, Tolwani A, Ronco C: Acute renal failure in critically ill patients: a multinational, multicenter study. JAMA 2005;294:813–818.

2 Bajwa A, Kinsey GR, Okusa MD: Immune mechanisms and novel pharmacological therapies of acute kidney injury. Curr Drug Targets 2009;10:1196–1204.

3 Himmelfarb J, Joannidis M, Molitoris B, Schietz M, Okusa MD, Warnock D, Laghi F, Goldstein SL, Prielipp R, Parikh CR, Pannu N, Lobo SM, Shah S, D'Intini V, Kellum JA: Evaluation and initial management of acute kidney injury. Clin J Am Soc Nephrol 2008;3:962–967.

4 Cruz DN, Bagshaw SM, Ronco C, Ricci Z: Acute kidney injury: classification and staging; in Ronco C, Costanzo MR, Bellomo R, Maisel AS (eds): Fluid Overload: Diagnosis and Management. Contrib Nephrol. Basel, Karger, 2010, vol 164, pp 24–32.

5 Amdur RL, Chawla LS, Amodeo S, Kimmel PL, Palant CE: Outcomes following diagnosis of acute renal failure in U.S. veterans: focus on acute tubular necrosis. Kidney Int 2009;76:1089–1097.

6 Ishani A, Xue JL, Himmelfarb J, Eggers PW, Kimmel PL, Molitoris BA, Collins AJ: Acute kidney injury increases risk of ESRD among elderly. J Am Soc Nephrol 2009;20:223–228.

7 Lo LJ, Go AS, Chertow GM, McCulloch CE, Fan D, Ordonez JD, Hsu CY: Dialysis-requiring acute renal failure increases the risk of progressive chronic kidney disease. Kidney Int 2009;76:893–899.

8 Wald R, Quinn RR, Luo J, Li P, Scales DC, Mamdani MM, Ray JG: Chronic dialysis and death among survivors of acute kidney injury requiring dialysis. JAMA 2009;302:1179–1185.

9 Pelte CH, Chawla LS: Novel therapeutic targets for prevention and therapy of sepsis associated acute kidney injury. Curr Drug Targets 2009;10:1205–1211.

10 Bellomo R, Cass A, Cole L, Finfer S, Gallagher M, Lo S, McArthur C, McGuinness S, Myburgh J, Norton R, Scheinkestel C, Su S: Intensity of continuous renal-replacement therapy in critically ill patients. N Engl J Med 2009:361:1627–1638.

11 Palevsky PM, Zhang JH, O'Connor TZ, Chertow GM, Crowley ST, Choudhury D, Finkel K, Kellum JA, Paganini E, Schein RM, Smith MW, Swanson KM, Thompson BT, Vijayan A, Watnick S, Star RA, Peduzzi P: Intensity of renal support in critically ill patients with acute kidney injury. N Engl J Med 2008;359:7–20.

12 Liano F, Felipe C, Tenorio MT, Rivera M, Abraira V, Saez-de-Urturi JM, Ocana J, Fuentes C, Severiano S: Long-term outcome of acute tubular necrosis: a contribution to its natural history. Kidney Int 2007;71:679–686.

13 Ishani A, Nelson D, Clothier B, Schult T, Nugent S, Greer N, Slinin Y, Ensrud KE: The magnitude of acute serum creatinine increase after cardiac surgery and the risk of chronic kidney disease, progression of kidney disease, and death. Arch Intern Med 2011:171:226–233.

14 Chawla LS, Amdur RL, Amodeo S, Kimmel PL, Palant CE: The severity of acute kidney injury predicts progression to chronic kidney disease. Kidney Int 2011; E-pub ahead of print.

15 Engelman DT, Adams DH, Byrne JG, Aranki SF, Collins JJ Jr, Couper GS, Allred EN, Cohn LH, Rizzo RJ: Impact of body mass index and albumin on morbidity and mortality after cardiac surgery. J Thorac Cardiovasc Surg 1999;118:866–873.

16 Gariballa SE, Parker SG, Taub N, Castleden CM: Influence of nutritional status on clinical outcome after acute stroke. Am J Clin Nutr 1998;68:275–281.

17 Owen WF Jr, Lew NL, Liu Y, Lowrie EG, Lazarus JM: The urea reduction ratio and serum albumin concentration as predictors of mortality in patients undergoing hemodialysis. N Engl J Med 1993:329:1001–1006.
18 Friedman AN, Fadem SZ: Reassessment of albumin as a nutritional marker in kidney disease. J Am Soc Nephrol 2010;21:223–230.
19 Chawla LS, Seneff MG, Nelson DR, Williams M, Levy H, Kimmel PL, Macias WL: Elevated plasma concentrations of IL-6 and elevated APACHE II score predict acute kidney injury in patients with severe sepsis. Clin J Am Soc Nephrol 2007;2:22–30.
20 Liu KD, Glidden DV, Eisner MD, Parsons PE, Ware LB, Wheeler A, Korpak A, Thompson BT, Chertow GM, Matthay MA: Predictive and pathogenetic value of plasma biomarkers for acute kidney injury in patients with acute lung injury*. Crit Care Med 2007, E-pub ahead of print.
21 Murugan R, Karajala-Subramanyam V, Lee M, Yende S, Kong L, Carter M, Angus DC, Kellum JA: Acute kidney injury in non-severe pneumonia is associated with an increased immune response and lower survival. Kidney Int 2010;77:527–535
22 Sutton TA, Mang HE, Campos SB, Sandoval RM, Yoder MC, Molitoris BA: Injury of the renal microvascular endothelium alters barrier function after ischemia. Am J Physiol Renal Physiol 2003;285:F191–F198.
23 Yang L, Besschetnova TY, Brooks CR, Shah JV, Bonventre JV: Epithelial cell cycle arrest in G2/M mediates kidney fibrosis after injury. Nat Med 2010;16:535–543, 1p following 143.
24 USRDS. USRDS Annual Report 2009; in USRDS, Minneapolis, MN, 2009.
25 Daugherty SL, Ho PM, Spertus JA, Jones PG, Bach RG, Krumholz HM, Peterson ED, Rumsfeld JS, Masoudi FA: Association of early follow-up after acute myocardial infarction with higher rates of medication use. Arch Intern Med 2008;168:485–491, discussion 492.
26 Ali T, Khan I, Simpson W, Prescott G, Townend J, Smith W, Macleod A: Incidence and outcomes in acute kidney injury: a comprehensive population-based study. J Am Soc Nephrol 2007;18:1292–1298.
27 Chertow GM, Burdick E, Honour M, Bonventre JV, Bates DW: Acute kidney injury, mortality, length of stay, and costs in hospitalized patients. J Am Soc Nephrol 2005;16:3365–3370.
28 Hsu CY, Chertow GM, McCulloch CE, Fan D, Ordonez JD, Go AS: Clin J Am Soc Nephrol 2009;4:891–898.

Lakhmir S. Chawla, MD
Department of Anesthesiology and Critical Care Medicine
George Washington University Medical Center
900 23rd Street, NW, RILF, G-105, Washington DC, 20037 (USA)
Tel. +1 202 715 4570, E-Mail minkchawla@gmail.com

Kellum JA, Ronco C, Vincent J-L (eds): Controversies in Acute Kidney Injury.
Contrib Nephrol. Basel, Karger, 2011, vol 174, pp 191–199

Role of Microvesicles in Acute Kidney Injury

Giovanni Camussi[a] · Vincenzo Cantaluppi[a] · Maria Chiara Deregibus[a] · Emanuele Gatti[b,c] · Ciro Tetta[b–d]

[a]Department of Internal Medicine, Centre for Molecular Biotechnology and Centre for Research in Experimental Medicine (CeRMS), Torino, Italy; [b]International Research and Development, Fresenius Medical Care, Bad Homburg, Germany; [c]Danube University, Center for Biomedical Technology, Krems, Austria; [d]Sis-Ter, Palazzo Pignano, Italy

Abstract

The main function of microvesicles (MVs) is signaling through specific interactions with target cells and transferring gene products. Therefore, they may participate in physiological and pathological processes. Gaining further insights into the molecular specificity of MVs has allowed identifying the cellular source and may provide new diagnostic tools in the future. Indeed, an increasing body of evidence indicates that MVs may offer prognostic information in various diseases such as chronic inflammation, cardiovascular and renal diseases, pathological pregnancy, tumors, and sepsis. The presence of MVs in body fluids makes them readily accessible. Their number, cellular origin, composition and function can be dependent on the state of the disease. In sepsis for example, activated endothelial cells may shed MVs that might trigger leukocyte and monocyte production and release pro-oxidant and inflammatory mediators. MVs from platelets may trigger disseminated intravascular coagulopathy. MVs are no doubt also involved in modulating immunity and future studies will clarify their functional role in negatively modulating the cell response. In addition, the recognition of the signals delivered by MVs may open new therapeutic strategies. The removal of harmful MVs from plasma may be beneficial in pathological conditions where MVs deliver thrombogenic and inflammatory signals. On the other hand, MVs derived from stem cells may reprogram altered functions in target cells suggesting that they could be exploited in regenerative medicine to repair damaged tissues. We will discuss the role of stem cell-derived MVs in the repair of acute kidney injury.

First described as platelet derivatives of less than 0.1 μm with procoagulant activity [1], extracellular vesicles such as exosomes, apoptotic blebs, microvesicles (MVs), microparticles, prostasomes and prominosomes were recognized from a large number of cell types, e.g. epithelial, hematopoietic, immune, placental, tumor and stem cells, and fibroblasts [2–4]. For a long time, the investigation on MVs was merely descriptive as MVs were considered to be inert cellular debris. The vesicles frequently observed by electron microscopy in the interstitial space or in blood were considered as the consequence of cell damage or as the result of dynamic plasma membrane turnover [5]. Only recently, however, have studies assigned specific functions to MVs released in the microenvironment by various cell types and in biological fluids such as blood, urine and exudates. The role of MVs is now evoked in regenerative medicine (multiorgan development, cell survival, differentiation), immune system regulation and several diseases [6].

In this review, we will discuss how MVs should be seen as two sides of the same coin and their potential role in the complex scenario of sepsis, the systemic response to infection associated with the development of multiple organ failure (MOF) and high mortality rates.

Biogenesis of Microvesicles

There has been a rather confusing nomenclature and some clarification is in order. Two terms that identify two different vesicles based on their size and mode of their release from the cell of origin are most commonly used. Exosomes are stored as intraluminal vesicles within multivesicular bodies of the late endosome and are released when these multivesicular bodies fuse with the cell membrane. Exosomes have an endosomal origin [7]. They are released by exocytosis through a mechanism dependent on cytoskeleton activation and under the regulation of p53 protein [8]. While exosomes are a rather homogenous population with a size of 30–120 nm, shedding vesicles are usually larger with a size ranging from 100 nm to 1 μm. Shedding vesicles originate from the budding of small cytoplasmic protrusions followed by their detachment from the cell surface, a process dependent on calcium influx, calpain and cytoskeleton reorganization [9]. MVs expose on their surface large amounts of phosphatidylserine from the inner leaflet to the outer bilayer of the cell membrane and are enriched in proteins associated with membrane lipid rafts [10] by calcium-dependent mechanisms that modify the asymmetric phospholipid distribution of plasma membranes by activation of specific enzymes named flippase, floppase and scramblase and by inhibition of translocase [11]. Moreover, the intracellular pathways that activate the reorganization of the cytoskeleton induce the detachment of plasma membrane protrusions from the cortical actin.

The release of MVs occurs in cells in a resting state or upon activation by soluble agonists, physical or chemical stress such as oxidative stress as well as hypoxia, or shear stress [12]. Blood levels of MVs are increased in various diseases including acute coronary syndrome [13], chronic renal failure [14] and sepsis [15]. MVs are a source of phospholipids, a substrate for phospholipase A2, which facilitates platelet aggregation, inflammation and chemotaxis of platelets and/or leukocytes to the endothelium, thus triggering production of monocyte cytokines (IL-1β, IL-8 and TNF-α). Upon stimulation, the plasma membrane of endothelial cells and monocytes express phosphatidylserine and tissue factor, allowing activation of coagulation factors (factor VII and thrombin at the cell surface). It is also evident that MVs may downregulate the procoagulant activity in response to the modulatory expression of the endothelial protein C receptor that binds protein C. Future studies will have to clarify the role of MVs in the amplification as well as in the control of feedback loops.

Modes of Action of Microvesicles

MVs may influence the behavior of target cells in multiple ways, such as signaling complexes by direct stimulation of target cells by transferring receptors between cells, delivery of proteins within the target cells or by a horizontal transfer of genetic information.

The most recent discovery includes the horizontal transfer of genetic information by MVs. For years, epigenetic changes were frequently reported in co-culture conditions. To explain this phenomenon, the transfer of genetic information between cells was shown [16]. MVs derived from human endothelial progenitors (EPC) shuttle mRNA to endothelial cells via interaction with α_4- and β_1-integrins expressed on their surface, thus activating an angiogenic program [17]. The molecular analysis of mRNA indicated that MVs derived from EPC were the cargo of a specific subset of cellular mRNA, including mRNA associated with pathways relevant for angiogenesis such as the *PI3K/AKT* and endothelial nitric oxide synthase signaling pathways. Furthermore, protein expression and functional studies demonstrated that PI3K and endothelial nitric oxide synthase were upregulated in target cells after MV incorporation. More recently, we demonstrated that MVs derived from human stem cells may also deliver in vivo human mRNA to mouse cells, resulting in protein translation [18, 19].

Besides mRNA, MVs may transfer microRNAs (miRNA) to target cells [20]. Since miRNAs are naturally occurring regulators of protein translation, this observation opens the possibility that stem cells can alter the expression of genes in neighboring cells by transferring miRNAs contained in MVs. We recently characterized miRNA shuttled by MVs released by human adult mesenchymal stem cells (MSCs) [21]. Hierarchical clustering and similarity analysis of miRNAs showed that miRNA compartmentalization and secretion by MVs are both highly regulated processes. Gene ontology analysis of predicted and validated targets showed that the highly expressed miRNAs in MVs derived from MSCs may be involved in multiorgan development, cell survival, differentiation and immune system regulation. It has been suggested that transfer of genetic information by MVs play a pivotal role in stem cell plasticity and tissue regeneration [22].This mechanism possibly contributes to the paracrine action of stem cells in the repair of tissue injury [23].

Sepsis: The As Yet Unraveled Complexity of 'Soluble Factors'

Sepsis, septic shock and MOF occur either as a response to infection or in the context of a noninfectious systemic inflammatory reaction syndrome, and are the main causes of mortality in intensive care units. The heart and the kidney are often acutely involved. The pathophysiology of the cellular and humoral alterations is complex and seems to be related not only to the ischemic response to hypoperfusion, but also to direct detrimental activity induced by circulating mediators with both pro- and anti-inflammatory properties able to interact in a dynamic manner and induce MOF. Sepsis-induced cardiac dysfunction has been known for many years, but the mechanism appears to be complex, including both 'intrinsic' cardiomyopathy and direct and/or indirect effects of circulating depressing factors [24].

Acute kidney injury (AKI) occurs very frequently in burn patients with MOF. We demonstrated that plasma derived from septic patients with severe burns induced apoptosis and functional alterations of glomerular podocytes and tubular epithelial cells [25]. Plasma collected from septic patients induced granulocyte adhesion, apoptosis and altered polarity in tubular cells – all biological events correlated to the onset of AKI. By using different adsorptive matrixes, the unselective removal of circulating 'soluble mediators' and MVs could limit plasma-induced tubular cell injury.

We recently demonstrated that adsorption significantly decreased plasma-induced tubular alterations and abated the concentrations of several soluble mediators such as TNF-α, Fas ligand and the CD40 ligand [26]. The inhibition of granulocyte adhesion to tubular cells was associated with the downregulation of ICAM-1 and CD40. Adsorption inhibited tubular cell apoptosis induced by septic plasma by downregulating the activation of caspase-3, -8, -9 and of Fas/death receptor-mediated signaling pathways. The alteration of cell polarity, morphogenesis, protein reabsorption and the downregulation of the tight junction molecule ZO-1, sodium transporter NHE3, glucose transporter GLUT-2 and of endocytic receptor megalin (all induced by septic plasma) were significantly reduced by resin adsorption.

The role of MVs in sepsis is presently under close scrutiny. Much evidence exists to support the contention that MVs may trigger the proinflammatory and pro-oxidant activities. In an experimental model of septic peritonitis in the rat,

Mortaza et al. [27] inoculated MVs from septic rats in healthy ones and reproduced hemodynamic septic inflammatory patterns associated with oxidative and nitrosative stresses. Of interest, the authors recognized a certain phenotype of MVs capable of exhibiting such an activity. Meziani et al. [15] reviewed the role of MVs in the pathogenesis of sepsis through multiple ways. MVs regulate vascular tone and are potent vascular proinflammatory and procoagulant mediators.

Perspectives in Microvesicle-Targeted Therapies

In the context of severe shock and MOF, the removal of MVs from the blood or plasma has so far had no proof-of-concept. While standard hemodialysis techniques, e.g. hemofiltration or hemodiafiltration, are not suitable, plasmapheresis and adsorption modalities would be most appropriate. The recent availability of diagnostics capable of quantifying MVs of a specific cell origin could be useful in understanding the cells/organs where injury is occurring. This approach would overlap with what has been done in oncology. The mRNA from MVs is enriched and differentiated to obtain a result that is indicative of the condition of tissue or organ from which the MVs originated. In tumors, these diagnostics could identify and stage by differential analysis of one or more distinct mRNAs, optionally together with identification and analysis of a non-RNA component of the MVs (Microvesicle-Based Compositions and Methods, United States Patent Application 20100075315). Adsorbents in extracorporeal therapies work by either direct adsorption of a given molecule, retention of a given target molecule, or adsorption/retention of both the molecules of interest and their carrier molecules (i.e. α_2-macroglobulin).

Adsorbent technologies have been intensely studied and have made tremendous advances in the last 20–30 years. These advances have been driven to a great extent by new developments in chromatography. A potential focus for future biotechnological research could be the application of selective removal of MVs expressing specific biomarkers. Antibodies or synthetic ligands immobilized on stationary matrix could be selected to act as acceptors for various proteins: Fas ligand, MHC I, MHC II, CD44, placental alkaline phosphatase, TSG-101, MHC I-peptide complex, MHC II-peptide complexes and in general any protein found to be expressed by MVs contributing to proinflammatory and/or anti-inflammatory biological activities.

Role of Stem Cell-Derived Microvesicles in the Repair of Acute Kidney Injury

The mechanism of stem cell-induced kidney repair after AKI is still controversial. The engraftment of stem cells derived from bone marrow and fusion are

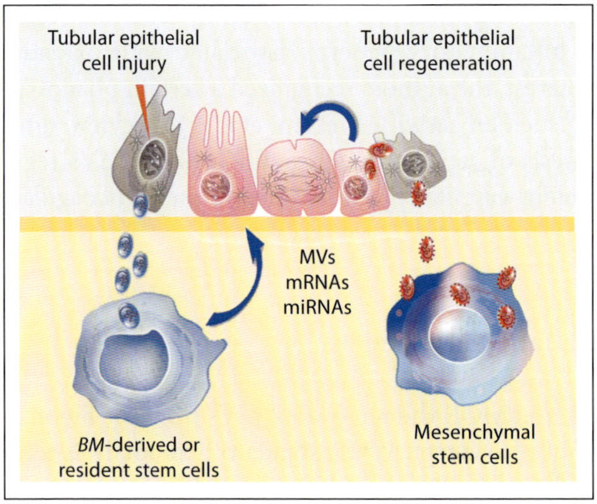

Fig. 1. Schematic representation of MV-mediated bidirectional exchange of genetic information between stem cells and tissue-injured cells. MVs released from injured tubular cells may reprogram the phenotype of bone morrow (BM)-derived or resident stem cells to acquire tissue specific phenotype by delivering the mRNAs and/or miRNA specifics of tubular cells to stem cells. On the other hand, MVs produced by stem cells, either resident or recruited from the circulation, or MVs produced ex vivo by exogenous stem cells and administered to experimental animals with AKI, may reprogram tubular cells which survived injury by delivering mRNA and/or miRNA that induce dedifferentiation, production of soluble paracrine mediators and cell cycle re-entry allowing tissue regeneration.

considered rare events and several studies suggest the involvement of a paracrine/endocrine mechanism rather than cell transdifferentiation. As an alternative to transdifferentiation and fusion, Quesenberry et al. [22] suggested that stem cell differentiation depends on epigenetic cell changes mediated by signals received from injured cells and delivered by MVs. They demonstrated that bone marrow cells co-cultured with injured lung cells induced the expression of lung-specific genes and proteins mediated by transfer of lung-specific mRNA delivered by MVs released from injured cells to bone marrow cells. In preliminary experiments, we found that MVs released from injured renal tubular epithelial cells may induce expression of tubular cell markers in human MSCs (fig. 1). Therefore, MVs released from injured tissue could reprogram bone marrow-derived stem cells as well as resident stem cells. MV-mediated transfer of genetic information could explain not only the plasticity and phenotypic changes of stem cells, but also the functional effects without the need of their transdifferentiation into tissue cells [28].

On the other hand, MVs derived from stem cells may reprogram cells which survived injury and favor tissue regeneration. It has been shown that MSC administration in AKI favors functional and morphological recovery

[28]. These beneficial effects are associated only with a transient recruitment of MSCs within the renal vasculature with a minimal engraftment within tubules [28]. Several studies on tubular repopulation after AKI indicate a prominent contribution of renal tubular cells which survived injury. Therefore, it has been suggested that MSC may provide paracrine support to the repair of injured tissue by releasing factors that limit injury and favor regeneration [23, 29].

We recently demonstrated that MVs released from stem cells recruited at the site of tissue injury may account for such paracrine/endocrine effects by inducing dedifferentiation of resident cells which survived injury and a tissue regenerative program (fig. 1) [23, 29]. MVs released from MSCs are able to deliver their mRNA cargo to the epithelial cells, thus stimulating proliferation and apoptosis resistance and accelerating functional and morphological recovery [18]. Given that the efficacy of MVs is comparable to that of MSC administration, we suggest that the beneficial effect of MSCs is mostly due to the release of MVs. The mechanism of repair is mainly dependent on RNA delivery as suggested by the inhibitory effect of RNA inactivation and by the presence in MVs of a defined subset of transcripts with multiple differentiative and functional properties of MSCs. Moreover, MSC-derived MVs may deliver miRNAs that activate translational control mechanisms or specific checkpoints for the transcripts after entry in target cells [21]. Therefore, administration of MVs produced by stem cells may provide a potential therapeutic strategy that would avoid the possible maldifferentiation of stem cells once engrafted in the kidney in the long term.

Conclusion

The main function of MVs is signaling through specific interactions with target cells and transferring gene products. Therefore, they may participate in physiological and pathological processes. Gaining further insights into the molecular specificity of MVs may allow for the identification of the cellular source and may provide new diagnostic tools. Indeed, an increasing body of evidence indicates that MVs may offer prognostic information in various diseases such as chronic inflammation, cardiovascular and renal diseases, pathological pregnancy, tumors, and sepsis. Many points require further investigation: (1) the stimuli and the molecular pathways that regulate the assembly within MVs of the biological active molecules that they shuttle, (2) the stimuli that trigger their release, (3) the surface receptors that may confer selective specificity, (4) the full diagnostic potential of MVs in different pathological conditions, (5) the strategy to inhibit formation or remove potentially harmful MVs from circulation, and (6) the therapeutic exploitation in regenerative medicine of the ability of MVs to modify the phenotype and function of target cells. The recognition of MVs has opened a new era, with new perspectives of investigation on the horizon.

Kellum JA, Ronco C, Vincent J-L (eds): Controversies in Acute Kidney Injury.
Contrib Nephrol. Basel, Karger, 2011, vol 174, pp 200–211

Controversies in Acute Kidney Injury: Effects of Fluid Overload on Outcome

Ravindra L. Mehta[a] · Josée Bouchard[b]

[a]Division of Nephrology and Hypertension, Department of Medicine, University of California San Diego, San Diego, Calif., USA; [b]Division of Nephrology, Department of Medicine, Université de Montréal, Montréal, Canada

Abstract

Fluid management in critically ill patients is a complex process as aggressive fluid resuscitation is commonly utilized for initial hemodynamic support and fluid administration often contributes to fluid retention, particularly when there is impaired kidney function. Recent evidence suggests that fluid accumulation is associated with adverse outcomes. It is unclear whether fluid retention is simply a marker of the severity of organ failure or a mediator of events. In this article we review the evidence and provide a framework for future studies to refine these concepts further.

Volume support is frequently required in critically ill patients exhibiting hypovolemia, particularly in the setting of shock, systemic inflammatory response syndrome and sepsis. Achieving an appropriate level of volume management requires knowledge of the underlying pathophysiology, evaluation of volume status, selection of an appropriate solution for volume repletion, and maintenance and modulation of the tissue perfusion and cellular injury. In the presence of a failing kidney, impaired fluid handling is common, but may not be well recognized. Consequently, volume management often results in a fluid overloaded state requiring diuresis or dialytic intervention. Over the last few years there is increasing recognition that fluid accumulation is associated with adverse outcomes including an increased risk for mortality. However, it is unclear whether fluid accumulation is simply a reflection of the underlying severity of organ dysfunction predisposing to fluid accumulation or an additional comorbidity, or represents a more sinister finding of mediating the events directly (fig. 1). This is a major area of controversy and offers opportunity for additional research. This review focuses on the consequences of a 'liberal' compared to a 'conservative or

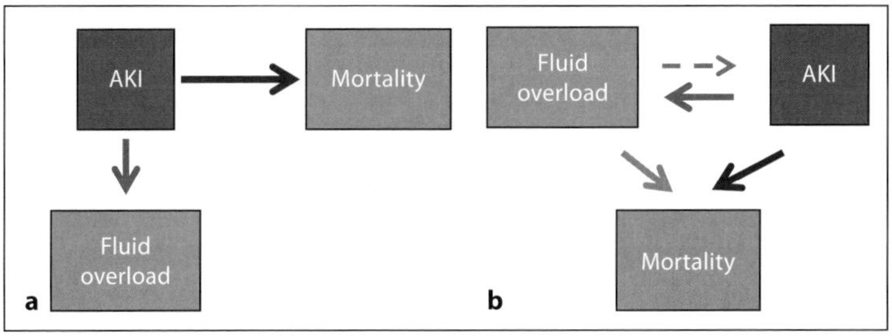

Fig. 1. Possible interactions of fluid overload and AKI. **a** Fluid accumulation is a consequence of AKI and may be a marker of the severity of renal dysfunction. **b** Fluid overload may occur independently of AKI, but could contribute to development of AKI and the consequent effects on mortality.

restrictive' fluid strategy on outcomes in critically ill patients, and whether fluid overload is a confounder or a direct mediator for adverse outcomes and how to address this question systematically in future research studies.

Definitions of Fluid Balance, Cumulative Fluid Balance, Fluid Overload and Fluid Administration Strategies

Fluid management in critically ill patients requires an accurate documentation of intakes and outputs; however, there is wide variation in how this information is recorded, reviewed and utilized for managing patients. We propose the following definitions to standardize the approach and facilitate comparisons. 'Daily fluid balance' refers to the daily difference in all intakes and outputs, which usually does not include insensible losses. 'Cumulative fluid balance' is the sum of daily fluid balance over a defined period of time. In contrast, the definition of 'fluid overload' varies widely among research publications, but usually implies a degree of pulmonary edema or peripheral edema. In this review, 'fluid accumulation' specifies conditions when there is a positive fluid balance, with or without associated fluid overload. The 'percentage of fluid overload adjusted for body weight' is the cumulative fluid balance expressed as a percent of body weight at baseline, usually at the admission in the intensive care unit [1, 2]. In previous studies, a cutoff of 10% or more has been associated with decreased survival [3, 4]. The definition of a 'liberal strategy' for fluid management has not been consistent in the past, but most often implies that fluid administration is not limited by consideration of fluid accumulation. The definition of a 'restrictive or conservative strategy' is usually also broad [5], but implies that fluid administration is somewhat restricted due to consideration of fluid accumulation. In the largest randomized controlled trial (RCT) on fluid administration in critically ill patients [6], the fluid strategy was well defined and fluid was administered depending on the presence or absence of shock, the requirement for vasopressors, and other parameters which included urine output, cardiac index, central venous pressure and pulmonary-artery occlusion pressure. This definition could be used as a framework to further clarify the terms 'liberal and conservative strategies' for future studies on fluid administration.

Relationship of Fluid Overload to Outcomes

Until recently, a 'liberal strategy' of fluid administration was most often used to avoid complications attributed to hypovolemia, such as hypotension, tachycardia, acute kidney injury (AKI), shock and multiorgan failure [7, 8]. In contrast, little attention focused on the consequences of fluid accumulation such as peripheral, tissue and organ edema; respiratory failure; hypertension; increased cardiac demand; and poor wound healing and delayed bowel recovery in the perioperative period [7, 8]. The recognition of these complications has led to the concept of a 'conservative strategy' for fluid management, defined in the previous section. Unfortunately, since the definitions of the liberal and conservative approaches have never reached a consensus, they vary in intensity and timing, which complicates their formal comparison. It is helpful to consider emerging evidence in this field in the context of different patient populations as discussed below.

Critically Ill Populations

In several clinical settings, such as acute respiratory distress syndrome, acute lung injury, sepsis, acute pulmonary edema, AKI and in the perioperative period, fluid overload has been associated with adverse outcomes, including increased mortality, prolonged duration of mechanical ventilation and impaired wound healing [2–4, 6, 9–13] (table 1).

Two RCTs have assessed the effect of a liberal versus a conservative approach to fluid management [6, 13]. In the first RCT, 172 patients undergoing colorectal surgery were randomized to a standard or a restrictive regimen of fluid management [13]. The quantity of fluid administered on the day of surgery [5,388 ml (range: 2,700–11,083) vs. 2,740 ml (range: 1,100–8,050), respectively; $p < 0.0005$] and changes in body weight from the day of the surgery to the end of measurement were significantly higher in the standard group ($p < 0.001$). This group presented decreased wound healing and increased postoperative cardiopulmonary complications compared to the restrictive one which were dose-related [13]. There was no difference in the survival rate or creatinine and blood urea nitrogen values between the two regimens throughout the study; however, the trial was not powered to evaluate changes in mortality.

In a trial of acute respiratory distress syndrome, Wiedemann et al. [6] randomized 1,000 patients to either a conservative or a liberal strategy of fluid

Table 1. Summary of studies assessing the effect of fluid overload on outcomes in critically ill patients

Author, year	Number of patients	Population	Study design	Intervention	Significant results
Simmons et al. [33] 1987	113	ARDS	prospective cohort	none	higher cumulative fluid balance and weight gain associated with increased mortality
Schuller et al. [10] 1991	89	pulmonary edema	retrospective cohort	none	higher fluid balance associated with increased mortality, length of hospitalization and days on mechanical ventilation
Goldstein et al. [1] 2001	21	pediatric AKI	retrospective cohort	none	higher fluid balance associated with mortality
Brandstrup et al. [13] 2003	172	elective colorectal surgery	RCT	restrictive vs. standard perioperative fluid strategy	restrictive strategy reduced cardiopulmonary and tissue-healing complications
Foland et al. [11] 2004	113	pediatric AKI	retrospective cohort	none	higher fluid balance associated with mortality
Gillespie et al. [3] 2004	77	pediatric AKI	retrospective cohort	none	higher fluid balance associated with mortality
Michael et al. [34] 2004	26	pediatric AKI	retrospective interventional	percentage of fluid overload <10%	100% of survivors vs. 40% of nonsurvivors had a percentage of fluid <10%
Goldstein et al. [2] 2005	116	pediatric AKI	prospective cohort	none	higher fluid balance associated with mortality
Sakr et al. [35] 2005	393	ALI/ARDS	prospective cohort	none	higher fluid balance associated with mortality
Uchino et al. [36] 2006	331	critically ill	prospective	none	higher fluid balance associated with mortality
Wiedemann et al. [6] 2006	1,000	ARDS	RCT	conservative vs. liberal fluid strategy	conservative strategy improved lung function and shortened the duration of mechanical ventilation
Payen et al. [15] 2008	1,120	critically ill	prospective cohort	none	higher fluid balance associated with mortality
Bouchard et al. [4] 2009	618	AKI	prospective cohort	none	higher fluid balance associated with mortality and possibly reduced renal recovery

Table 1. Continued

Author, year	Number of patients	Population	Study design	Intervention	Significant results
Sutherland et al. [12] 2010	297	AKI	prospective cohort	none	higher fluid balance associated with mortality
Boyd et al. [14] 2011	778	septic shock VASST study	retrospective analysis of RCT	none	a more positive fluid balance at both at 12 h and day 4 correlated significantly with increased mortality
Grams et al. [16] 2011	306	subset of ARDS patients in FACTT trial with AKI	retrospective analysis of RCT	conservative vs. liberal fluid strategy	a positive fluid balance after AKI was strongly associated with mortality; diuretic use did not influence outcome

ALI = Acute lung injury; ARDS = acute respiratory distress syndrome; VASST = Vasopressin in Septic Shock Trial.

management. The mean (±SE) cumulative fluid balance during the first 7 days was –136 ± 491 ml in the conservative group and 6,992 ± 502 ml in the liberal group ($p < 0.001$). They did not report body weights in the study. The primary outcome, mortality rate at 60 days, was similar between the two groups and the duration of mechanical ventilation was significantly improved with the conservative strategy. Although the incidence of AKI was not reported, there were no significant differences in mean creatinine values over the first 7 days of the study.

More recently, Boyd et al. [14] reviewed the fluid balance data in the first 4 days of the Vasopressin in Septic Shock Trial, which enrolled 778 patients who had septic shock and who were receiving a minimum of 5 μg of norepinephrine per minute. There was a significant increase in cumulative fluid balance within 12 h of presentation of +4.2 liters to +11 liters at 4 days. After correcting for age and severity of illness, a more positive fluid balance at both 12 h and day 4 correlated significantly with increased mortality. While initial vales of central venous pressure correlated with fluid balance, values later in the course were not associated.

AKI Populations

In AKI, four multicenter prospective observational studies have shown a relationship between fluid overload and mortality: two in pediatric and two in adult patients [2, 4, 12, 15]. In the first pediatric study, the percentage of fluid overload adjusted for body weight at continuous renal replacement therapy (CRRT) initiation was significantly higher in nonsurvivors versus survivors (25.4 ± 32.9

vs. 14.2 ± 15.9%; $p < 0.03$), even after adjustment for severity of illness [2]. In the most recent study [12], the adjusted mortality OR was 1.03 (95% CI: 1.01–1.05), which suggests a 3% increase in mortality for each 1% increase in severity of fluid overload. When fluid overload was dichotomized to ≥20% and <20%, patients with ≥20% fluid overload had an adjusted mortality OR of 8.5 (95% CI: 2.8–25.7).

In 1,120 adults with sepsis-induced AKI, fluid accumulation was also associated with decreased survival [15]. In this study, patients had severe AKI, but did not necessarily require dialysis. The mean daily fluid balance was 0.15 ± 1.06 l/day in survivors versus 0.98 ± 1.5 l/day in nonsurvivors ($p < 0.0001$). The association between fluid overload and mortality remained significant only in patients with early AKI, defined as AKI occurring within 2 days of intensive care unit admission. In the study performed by the PICARD group, patients with fluid overload, defined as a percentage of fluid accumulation >10% from admission body weight, had a significant higher mortality at 60 days (46 vs. 32%; $p = 0.006$) [4]. Among dialyzed patients, survivors had a lower percentage of fluid accumulation at dialysis initiation compared with nonsurvivors [8.8 vs. 14.2%; $p = 0.01$ adjusted for Acute Physiology and Chronic Health Evaluation (APACHE) III score and dialysis modality]. In patients who did not require dialysis, survivors also presented less fluid accumulation adjusted for body weight at peak creatinine (4.5 vs. 10.1%; $p = 0.03$ adjusted for APACHE III score). Patients with fluid overload at peak creatinine were also less likely to recover kidney function (35 vs. 52%; $p = 0.007$); however, there were inconsistent results in the relationship between fluid overload at dialysis initiation and dialysis independence at hospital discharge (41 vs. 32% in fluid-overloaded patients; $p = 0.21$) [4]. In any case, these findings are in opposition to the common belief that fluid accumulation may protect the kidneys. A recent analysis of the FACTT trial data [16] limited to patients who had AKI showed that higher fluid balance post-AKI was significantly associated with mortality in both crude and adjusted analysis. While diuretic use post-AKI had a protective effect on mortality, this was nonsignificant after adjustment for post-AKI fluid balance.

Is Fluid Overload a Marker of Disease Severity?

All four previously mentioned studies in AKI [2, 4, 12, 15] have adjusted the association between fluid overload and mortality with a severity of illness score: the Pediatric Risk of Mortality (PRISM 2) score in pediatric patients and APACHE III and Simplified Acute Physiology Score (SAPS II) score in adult patients, respectively. Therefore, even if fluid overload is a marker of disease severity, all results from recent prospective studies in AKI have been adjusted to take into account this potential confounding effect. However, there might still be some other unknown factors which have not been included in the analyses.

In addition, in two randomized controlled studies, fluid overload independently increased complications, further refuting the hypothesis that fluid accumulation is simply a marker of disease severity [6, 13].

Is Fluid Overload a Mediator of Outcome?

The mechanisms by which fluid overload could influence outcomes have never been clearly demonstrated. The 'third space' most probably accumulates in the interstitium due to the destruction of an integral structure of the vascular wall, the endothelial glycocalyx [17]. The endothelial glycocalyx seems to retain proteins hydrostatically forced out of the vascular wall. In experimental studies, ischemia/reperfusion, TNF-α and atrial natriuretic peptide could all lead to the destruction of this structure [18, 19]. Since acute hypervolemia triggers atrial natriuretic peptide release, avoiding intravascular hypervolemia could theoretically protect the endothelial glycocalyx and therefore prevent protein-rich plasma shifts and subsequent interstitial edema. Additionally, it appears that further studies on the mechanisms underlying fluid shifts between compartments could help our understanding on the effect of fluid overload on the vasculature and cells.

There are limited data to determine whether significant fluid accumulation can contribute to AKI either directly or through aggressive diuretic use; however, several studies, including RCTs, have supported the role of fluid accumulation on adverse outcomes in critically ill patients, as previously discussed [6, 13].

One of the most evident examples of fluid retention contributing to AKI comes from the glomerular literature. In minimal change nephrotic syndrome, kidney dysfunction can occur due to fluid retention and associated interstitial edema and compression of tubules [20]. Hypovolemia due to concomitant hypoalbuminemia and aggressive diuretic therapy can also lead to AKI; however, this complication is more common when serum albumin is less than 1.5 g/dl, a situation that rarely occurs in AKI without associated glomerular injury [21]. Other examples of fluid overload contributing to AKI might include congestive heart failure (CHF), multiple organ dysfunction syndrome and the use of diuretics.

In CHF, due to reduced cardiac output, there are concomitant prerenal azotemia and edema formation secondary to salt retention [22]. Whether edema during CHF can directly contribute to AKI remains unknown. Interestingly, in chronic CHF, poor forward flow alone does not seem to have an influence on the development of chronic kidney disease or worsening of kidney function [23].

In multiple organ dysfunction syndrome, edema is common and can occur during severe sepsis – with or without associated AKI – due to the release of complement factors, cytokines, prostaglandin products and altered organ microcirculation [21]. In this setting, edema is attributed to a combination of increased capillary permeability to proteins and increased net transcapillary hydrostatic pressure through reduced precapillary vasoconstriction. To our knowledge, the relationship between peripheral edema and kidney function has not been carefully studied. The use of positive end-expiratory pressure (PEEP) in these patients can also exacerbate fluid and salt retention [24], and decrease lymphatic drainage [25]. In addition, several studies have shown that mechanical ventilation and PEEP reduce urine output; however, their effects on glomerular filtration rate and renal blood flow are inconsistent and may reflect differences in hydration status and lung injury [26].

In summary, there are no clear indications that fluid overload per se can contribute to AKI. On the other hand, fluid overload increases the duration of mechanical ventilation, cardiopulmonary complications and the healing process following surgery, possibly through impaired diffusive tissular transport of nutrients and removal of waste from the tissues. The type of fluid (colloids vs. crystalloids) has also been associated with development of AKI. However, to our knowledge only a few studies performed on the association between the type of fluid administered and AKI have included the overall level of fluid accumulation in relation to the outcomes [27]. This is a potential confounder for the association and further studies are required to address these questions.

How Does Fluid Overload Cause Organ Dysfunction?

Fluid overload presenting as peripheral edema has been conceptualized as an imbalance between net transcapillary fluid filtration and fluid removal from the tissues by the lymphatic system [21]. Therefore, edema can occur secondary to an increase in net transcapillary hydrostatic pressure (e.g. in CHF) or capillary permeability to proteins (e.g. in sepsis), and to a decrease in net transcapillary oncotic pressure (e.g. in nephrotic syndrome) or lymphatic drainage (e.g. PEEP ventilation).

Fluid accumulation can contribute to impair organ function by different mechanisms. First, tissue edema can directly impair gut absorption or kidney excretion. Fluid accumulation can also lead to increased abdominal pressure and/or renal venous congestion. The abdominal compartment syndrome has been increasingly recognized in surgical and medical patients following intense volume resuscitation, with consequent acute formation of ascites and visceral edema [28]. The abdominal compartment syndrome has been directly involved in several organ dysfunctions, such as renal and respiratory failures and intestinal and hepatic ischemia [29]. With regards to renal venous congestion, one study in chronic CHF has shown that among several hemodynamic parameters, only the right atrial pressure correlated weakly with serum creatinine ($r = 0.165$, $p = 0.03$) [23]. These results suggest that renal congestion may have

5 Varadhan KK, Lobo DN: A meta-analysis of randomised controlled trials of intravenous fluid therapy in major elective open abdominal surgery: getting the balance right. Proc Nutr Soc 2010;69:488–498.
6 Wiedemann HP, Wheeler AP, Bernard GR, Thompson BT, Hayden D, deBoisblanc B, Connors AF Jr, Hite RD, Harabin AL: Comparison of two fluid-management strategies in acute lung injury. N Engl J Med 2006;354:2564–2575.
7 Hilton AK, Pellegrino VA, Scheinkestel CD: Avoiding common problems associated with intravenous fluid therapy. Med J Aust 2008;189:509–513.
8 Bundgaard-Nielsen M, Secher NH, Kehlet H: 'Liberal' vs. 'restrictive' perioperative fluid therapy – a critical assessment of the evidence. Acta Anaesthesiol Scand 2009;53:843–851.
9 Vincent JL, Sakr Y, Sprung CL, Ranieri VM, Reinhart K, Gerlach H, Moreno R, Carlet J, Le Gall JR, Payen D: Sepsis in European intensive care units: results of the SOAP study. Crit Care Med 2006;34:344–353.
10 Schuller D, Mitchell JP, Calandrino FS, Schuster DP: Fluid balance during pulmonary edema. Is fluid gain a marker or a cause of poor outcome? Chest 1991;100:1068–1075.
11 Foland JA, Fortenberry JD, Warshaw BL, Pettignano R, Merritt RK, Heard ML, Rogers K, Reid C, Tanner AJ, Easley KA: Fluid overload before continuous hemofiltration and survival in critically ill children: a retrospective analysis. Crit Care Med 2004;32:1771–1776.
12 Sutherland SM, Zappitelli M, Alexander SR, Chua AN, Brophy PD, Bunchman TE, Hackbarth R, Somers MJ, Baum M, Symons JM, Flores FX, Benfield M, Askenazi D, Chand D, Fortenberry JD, Mahan JD, McBryde K, Blowey D, Goldstein SL: Fluid overload and mortality in children receiving continuous renal replacement therapy: the prospective pediatric continuous renal replacement therapy registry. Am J Kidney Dis 2010;55:316–325.
13 Brandstrup B, Tonnesen H, Beier-Holgersen R, Hjortso E, Ording H, Lindorff-Larsen K, Rasmussen MS, Lanng C, Wallin L, Iversen LH, Gramkow CS, Okholm M, Blemmer T, Svendsen PE, Rottensten HH, Thage B, Riis J, Jeppesen IS, Teilum D, Christensen AM, Graungaard B, Pott F: Effects of intravenous fluid restriction on postoperative complications: comparison of two perioperative fluid regimens: a randomized assessor-blinded multicenter trial. Ann Surg 2003;238:641–648.
14 Boyd JH, Forbes J, Nakada TA, Walley KR, Russell JA: Fluid resuscitation in septic shock: a positive fluid balance and elevated central venous pressure are associated with increased mortality. Crit Care Med 2011;39:259–265.
15 Payen D, de Pont AC, Sakr Y, Spies C, Reinhart K, Vincent JL: A positive fluid balance is associated with a worse outcome in patients with acute renal failure. Crit Care 2008;12:R74.
16 Grams ME, Estrella MM, Coresh J, Brower RG, Liu KD: Fluid balance, diuretic use, and mortality in acute kidney injury. Clin J Am Soc Nephrol 2011;6:966–973.
17 Jacob M, Chappell D, Rehm M: The 'third space' – fact or fiction? Best Pract Res Clin Anaesthesiol 2009;23:145–157.
18 Chappell D, Dorfler N, Jacob M, Rehm M, Welsch U, Conzen P, Becker BF: Glycocalyx protection reduces leukocyte adhesion after ischemia/reperfusion. Shock 2010;34:133–139.
19 Becker BF, Chappell D, Bruegger D, Annecke T, Jacob M: Therapeutic strategies targeting the endothelial glycocalyx: acute deficits, but great potential. Cardiovasc Res 2010;87:300–310.
20 Lowenstein J, Schacht RG, Baldwin DS: Renal failure in minimal change nephrotic syndrome. Am J Med 1981;70:227–233.
21 Andreucci M, Federico S, Andreucci VE: Edema and acute renal failure. Semin Nephrol 2001;21:251–256.
22 Ronco C, Haapio M, House AA, Anavekar N, Bellomo R: Cardiorenal syndrome. J Am Coll Cardiol 2008;52:1527–1539.

23 Nohria A, Hasselblad V, Stebbins A, Pauly DF, Fonarow GC, Shah M, Yancy CW, Califf RM, Stevenson LW, Hill JA: Cardiorenal interactions: insights from the ESCAPE trial. J Am Coll Cardiol 2008;51:1268–1274.

24 Andrivet P, Adnot S, Brun-Buisson C, Chabrier PE, Darmon JY, Braquet P, Lemaire F: Involvement of ANF in the acute antidiuresis during PEEP ventilation. J Appl Physiol 1988;65:1967–1974.

25 Gest AL, Bair DK, Vander Straten MC: The effect of outflow pressure upon thoracic duct lymph flow rate in fetal sheep. Pediatr Res 1992;32:585–588.

26 Pannu N, Mehta RL: Effect of mechanical ventilation on the kidney. Best Pract Res Clin Anaesthesiol 2004;18:189–203.

27 Myburgh J, Cooper J, Finfer S, Bellomo R, Norton R, Bishop N, Kai Lo S, Vallance S: Saline or albumin for fluid resuscitation in patients with traumatic brain injury. N Engl J Med 2007;357:874–884.

28 Maerz L, Kaplan LJ: Abdominal compartment syndrome. Crit Care Med 2008;36:S212–S215.

29 Saggi BH, Sugerman HJ, Ivatury RR, Bloomfield GL: Abdominal compartment syndrome. J Trauma 1998;45:597–609.

30 Marjanovic G, Villain C, Juettner E, zur Hausen A, Hoeppner J, Hopt UT, Drognitz O, Obermaier R: Impact of different crystalloid volume regimes on intestinal anastomotic stability. Ann Surg 2009;249:181–185.

31 Berthiaume Y, Matthay MA: Alveolar edema fluid clearance and acute lung injury. Respir Physiol Neurobiol 2007;159:350–359.

32 Lobo DN, Bostock KA, Neal KR, Perkins AC, Rowlands BJ, Allison SP: Effect of salt and water balance on recovery of gastrointestinal function after elective colonic resection: a randomised controlled trial. Lancet 2002;359:1812–1818.

33 Simmons RS, Berdine GG, Seidenfeld JJ, Prihoda TJ, Harris GD, Smith JD, Gilbert TJ, Mota E, Johanson WG Jr.: Fluid balance and the adult respiratory distress syndrome. Am Rev Respir Dis 1987;135:924–929.

34 Michael M, Kuehnle I, Goldstein SL: Fluid overload and acute renal failure in pediatric stem cell transplant patients. Pediatr Nephrol 2004;19:91–95.

35 Sakr Y, Vincent JL, Reinhart K, Groeneveld J, Michalopoulos A, Sprung CL, Artigas A, Ranieri VM: High tidal volume and positive fluid balance are associated with worse outcome in acute lung injury. Chest 2005;128:3098–3108.

36 Uchino S, Bellomo R, Morimatsu H, Sugihara M, French C, Stephens D, Wendon J, Honore P, Mulder J, Turner A: Pulmonary artery catheter versus pulse contour analysis: a prospective epidemiological study. Crit Care 2006;10:R174.

Ravindra L. Mehta, MD
200 W Arbor Drive
Mail Code 8342
San Diego, CA 92103 (USA)
Tel. +1 619 543 7310, E-Mail rmehta@ucsd.edu

Kellum JA, Ronco C, Vincent J-L (eds): Controversies in Acute Kidney Injury.
Contrib Nephrol. Basel, Karger, 2011, vol 174, pp 212–221

Does Renal Replacement Therapy Improve Outcome? Controversies in Acute Kidney Injury

Patrick Murray[a] · Suneel Udani[b] · Jay L. Koyner[b]

[a]School of Medicine and Medical Science, University College Dublin, Dublin, Ireland;
[b]Section of Nephrology, Department of Medicine, University of Chicago, Chicago, Ill., USA

Abstract

All aspects of current treatment of acute kidney injury (AKI), including renal replacement therapy (RRT), are basically supportive. Emergent RRT is indicated in the management of AKI with refractory pulmonary edema, hyperkalemia or metabolic acidosis, or when uremic symptoms or signs develop. More aggressive practitioners use prophylactic RRT in patients with sustained anuria, persistent oliguria with progressive azotemia and a probable glomerular filtration rate <10 ml/min, or to prevent uncontrolled positive fluid balance in patients with AKI. However, this approach to RRT initiation in AKI is largely supported by retrospective analyses rather than prospective clinical trials. The approach to RRT dosing in AKI is more evidence-based. Outcomes in single-center studies of higher intensity versus standard RRT (intermittent and/or continuous) have been inconsistent. However, two large multicenter negative randomized trials have shifted the weight of evidence towards suggesting provision of an effectively delivered standard dose of RRT in AKI, rather than seeking to increase RRT intensity.

Despite decades of progress in the care of the critically ill, the development of AKI still portends a poor outcome, particularly if renal replacement therapy (RRT) is required. Progress in intensive care nephrology has included the development of continuous RRT (CRRT) modalities that can provide effective renal support for the control of fluid balance and uremia with less hemodynamic instability than intermittent hemodialysis (IHD), and an increasing evidence base guiding the dosing of RRT in AKI. Driven by the desire to increase RRT dose and improve fluid balance control, and facilitated by the availability of more hemodynamically tolerable continuous modalities [1, 2], nephrologists and intensivists have increasingly moved towards earlier RRT initiation in AKI. Similar to the management of other medical emergencies, such as the use of supplemental oxygen ± mechanical ventilation for refractory hypoxemia, or the use of vasopressors for vasodilatory hypotension that is refractory to adequate fluid resuscitation, in many cases the use of RRT is indicated by urgency and common sense, rather than controlled trials or meta-analyses.

There is a longstanding consensus that RRT should be initiated if there are uremic symptoms or signs, or if volume overload, electrolyte abnormalities or acidosis are refractory to medical management [2, 3]. Otherwise, the precise timing of RRT initiation is usually a matter of clinical judgment. Absent the above clinical indications, hemodialysis (HD) is commonly initiated when the blood urea nitrogen (BUN) concentration reaches 100 mg/dl, and repeated to maintain predialysis BUN below 80 mg/dl. This pattern of practice is based primarily on early experience which suggested uremic bleeding diathesis and hemorrhage were reduced when HD was initiated before BUN exceeded 100 mg/dl [3], the results of small prospective AKI studies in 1975 and 1986 [4, 5], and extrapolation from results of the National Cooperative Dialysis Study in chronic renal failure patients. This relatively prophylactic approach is balanced against potential adverse effects of RRT, including hypotension with worsening renal ischemic acute tubular necrosis, and the renal and systemic inflammatory effects of extracorporeal membrane exposure.

Although it seems prudent to start RRT aggressively in critically ill catabolic patients, the literature has not conclusively shown a significant benefit of prophylactic dialysis [2], including a recent negative small, prospective randomized trial using CRRT by Bouman et al. [6]. Recent studies that purport to demonstrate a benefit of early RRT are nonrandomized retrospective cohort studies, which generally divide populations into early or late initiation groups based upon BUN level [2, 7–11] (table 1), or in some studies by degree of fluid overload [12]. Although some of these data are compelling, particularly the fluid overload literature, they are also unavoidably confounded by design and supportive of the premise that there is equipoise for prospective trials of early versus late initiation of RRT in AKI, rather than conclusive evidence to guide this clinical decision. Meanwhile, in the absence of such definitive positive trials, it is universally accepted that emergent RRT is indicated in the management of AKI when refractory pulmonary edema, hyperkalemia or metabolic acidosis develop, or to treat symptomatic uremia. In the opinion of many practitioners, initiation of RRT should also be considered in patients with sustained anuria, persistent oliguria with progressive azotemia and glomerular filtration rate <10 ml/min, and to control fluid balance rather than waiting for complications of volume overload, electrolyte imbalance or uremia to develop. Such approaches, however, are not well supported by evidence.

Table 1. Studies of RRT initiation in AKI

Study	Year	Design	Number of patients	Early initiation criteria	Late initiation criteria	Recovery of renal function	Survival
Conger [4]	1975	RCT	18	BUN <70 mg/dl or SCr <5 mg/dl	BUN >150 mg/dl or SCr >10 mg/dl or clinical indication		early: 64%; late: 20%
Gillum et al. [5]	1986	RCT	34	treatment goal BUN <60 mg/dl and SCr <5 mg/dl	treatment goal BUN <100 mg/dl and SCr <9 mg/dl		early: 41%; late: 53%
Bouman et al. [6]	2002	RCT	106	<12 h after meeting definition for AKI requiring RRT	BUN >112 mg/dl, K >6.5 mmol/l or severe cardiogenic pulmonary edema		early high-dose: 74.3%; early low-dose: 68.6%; late low-dose: 75%
Gettings et al. [7]	1999	retrospective observational	100	BUN <60 mg/dl	BUN >60 mg/dl	early: 100%; late: 91.6%	early: 39%; late: 20%
Demirkilic et al. [8]	2004	retrospective observational	61	urine output <100 ml × 8 h despite diuretic	SCr 5 mg/dl or K >5.5 meq/l		early: 76.5%; late: 45.5%
Elahi et al. [9]	2004	retrospective observational	64	urine output 100 ml × 8 h despite diuretic	urea >84 mg/dl or SCr >3.39 mg/dl or K >6 meq/l		early: 78%; late: 57%
Wu et al. [10]	2007	retrospective observational	80	BUN <80 mg/dl	BUN >80 mg/dl	early: 39.2%; late: 12%	early: 37%; late: 15.4%
Liu et al. [11]	2006	retrospective observational	243	BUN ≤76 mg/dl	BUN >76 mg/dl		early: 65%; late: 59%

In contrast to the approach to the initiation of renal support, the prescription of acute RRT dosing has increasingly been guided by high-quality evidence in recent years (table 2). In patients receiving IHD, the approach to dosing has been adapted from that used for chronic RRT in patients with end-stage renal disease. Clearance of urea, a low-molecular-weight nitrogenous waste product, is the most commonly studied marker of adequate uremic detoxification by HD. The term Kt/V is a unitless measure of HD dose (based on urea removal): K is the urea clearance of the dialysis membrane used (in milliliters per minute), t is the duration of dialysis (in minutes), and V is the volume of distribution of urea in the patient (in milliliters). Thus, Kt/V is a measure of the volume of plasma cleared of urea during a HD session (Kt) divided by the urea distribution volume (V, assumed to be total body water: 0.5–0.6 l/kg), and larger Kt/V values signify greater HD dose. Kt/V measurements of delivered dialysis dose are usually calculated using the ratio of postdialysis to predialysis BUN and a nomogram. In chronic HD patients the minimum acceptable dialysis dose is a Kt/V of 1.2 per IHD treatment. Extrapolation of HD data from the chronic HD population to the AKI setting is limited by numerous poorly quantifiable variables, including effects of critical illness on urea kinetic modeling [13]. Furthermore, numerous patient- and technique-related factors impede effective delivery of prescribed dialysis in critically ill patients.

In a retrospective analysis by Paganini et al. [14], improved survival was seen in AKI patients with higher Kt/V, and this survival benefit was particularly evident in patients with moderate-range severity of illness. In patients with low or high severity of illness, mortality was not affected by dialysis dose. Subsequently, Schiffl et al. [15] found that patients who received daily IHD had lower mortality rates compared with patients who received alternate-day IHD (28 vs. 46%), and accelerated renal recovery. In patients who received alternate-day IHD, the time-averaged BUN level was 104 mg/dl, reflecting inadequate dialysis that is below the standard of care for most critically ill patients. Therefore, it is not surprising that their outcome was inferior to better-dialyzed patients. Nonetheless, these data suggested that the prudent approach is to prescribe more frequent dialysis for intensive care unit (ICU) patients treated with IHD, to optimize dose delivery, fluid balance and outcome.

The past decade has also seen the emergence of outcome data to guide CRRT dosing, initially from the landmark study by Ronco et al. [16], who conducted a large prospective trial to determine the impact of hemofiltration dose on survival. In this study, patients were randomly assigned to one of three treatment doses, based on effluent volume (which is generally equivalent to small solute clearance by continuous hemofiltration, or by low-volume continuous HD): 20 ml/kg/h (group 1), 35 ml/kg/h (group 2) or 45 ml/kg/h (group 3). Group 1 had a higher mortality rate than those in groups 2 and 3. Based on this study, during the past decade many practitioners prescribed a minimum effluent volume of 35 ml/kg/h for ICU patients with AKI treated with CRRT. However, the majority of subsequent multicenter randomized trials and meta-analyses have suggested that lower RRT doses may be adequate, if effectively delivered.

In another single center study, Tolwani et al. [17] randomized 200 patients to 'standard-dose' continuous venovenous hemodiafiltration (CVVHDF; prescribed 20 ml/kg/h) versus 'high-dose' (35 ml/kg/h), stratified by presence of sepsis and oliguria. They found no difference in survival or renal recovery between the groups. Specifically, the survival (to ICU discharge or 30 days)

Table 2. Studies of RRT dosing in AKI

Study	Year	Design	Number of patients	Conventional dose	High dose	Recovery of renal function	Survival
RENAL [22]	2009	multicenter RCT	1,508	CVVHDF, postfilter dilution, effluent rate: 25 ml/kg/h	CVVHDF, postfilter dilution, effluent rate: 40 ml/kg/h	intensified: 85.5%; standard: 87.3%	conventional dose: 45%; high dose: 45%
Faulhaber-Walter et al. [18]	2009	single center RCT	156	standard SLEDD, median delivered dialysate volume 4 l/kg	intensified SLEDD, median delivered dialysate volume 12 l/kg	intensified: 60%; standard: 63%	intensified SLEDD: 55.6%; standard SLEDD: 61.3%
ATN [20]	2008	multicenter RCT	1,124	CRRT-prefilter dilution, effluent rate: 20 ml/kg/h or IHD-3 times per/week	CRRT-prefilter dilution, effluent rate: 35 ml/kg/h or IHD-6 times per/week	intensive: 15.4%; standard: 18.4%	conventional dose: 48%; high dose: 46%
Tolwani et al. [17]	2008	single-center RCT	200	CVVHDF prefilter dilution, effluent rate: 20 ml/kg/h	CVVHDF prefilter dilution, effluent rate: 35 ml/kg/h	high-dose: 69%; standard dose: 80%	conventional dose: 56%; high dose: 49%
Saudan et al. [19]	2006	single-center RCT	204	CVVH, effluent rate: 1–2.5 l/h	CVVH, effluent rate: 1–2.5 l + HD 1–1.5 l/h	CVVH: 71%; CVVHDF: 78%	conventional dose: 39%; high dose: 59%
Bouman et al. [6]	2002	multicenter RCT	106	CVVH, effluent rate: 1–1.5 l/h	CVVH, effluent rate: 3–4 l/h		conventional dose: 68.6%; high dose: 74.3%
Ronco et al. [16]	2000	single-center RCT	425	CVVH postfilter dilution, effluent rate: 20 ml/kg/h	CVVH postfilter dilution, effluent rate: 35 or 45 ml/kg/h		conventional dose: 41%; 35 ml/kg/h: 57%; 45 ml/kg/h: 58%
Schiffl et al. [15]	2002	single-center RCT	146	IHD alternate day	IHD daily		conventional dose: 54%; high dose: 72%

was 56% in the standard dose group and 49% in the high-dose group (p = NS). Renal recovery was 41% in the standard-dose group and 29% in the high-dose group (p = NS). Similarly, in a single-center study of patients with AKI receiving diffusive therapy with SLEDD (sustained low efficiency daily dialysis), Faulhaber-Walter et al. [18] randomized 156 subjects to high-intensity versus standard SLEDD, and found no improvement in mortality or renal recovery. In contrast, Saudan et al. [19] randomized 206 patients to receive CRRT with either hemofiltration alone (1–2.5 l/h) versus a higher dose and achieved the addition of dialysate (1–1.5 l/h). As such, this trial was both a comparison of dose and modality (continuous venovenous hemofiltration vs. CVVHDF). They found markedly improved 28-day survival in the higher-dose CVVHDF group (59 vs. 39%, p = 0.03), but no difference in recovery of renal function (78 vs. 71%, p = 0.62). In summary, these single-center trials of RRT dosing were inconsistent in their outcome findings; subsequently, two large multicenter trials have provided more definitive results.

Palevsky et al. [20] conducted the VA/NIH ATN trial, which randomized the intensity of RRT dosing in AKI patients enrolled in 27 centers. This trial randomized patients to higher versus lower intensity RRT, which was delivered by IHD when the subjects were hemodynamically stable and CRRT during unstable periods. High-intensity RRT was either IHD prescribed 6 days weekly, or CVVHDF with a prescribed effluent rate of 35 ml/kg/h. Standard intensity RRT was either IHD 3 days per week, or CVVHDF with a prescribed effluent rate of 20 ml/kg/h. IHD treatments were prescribed with a Kt/V of 1.2–1.4, and additional isolated ultrafiltration sessions were permitted in the standard therapy arm for volume management. 1,124 patients were randomized: 563 to the intensive strategy and 561 to the standard strategy. Mortality at 60 days was the primary endpoint, which did not differ between the groups (53.6% for high-intensity therapy vs. 51.5% for standard therapy, p = NS). Similarly, 28-day renal recovery was not affected by RRT intensity (75.8% of individuals in the high-intensity group remained RRT-dependent or had only partial renal recovery compared to 72.6% in the standard therapy group, p = NS). Thus, an adequately powered, multicenter trial that fully accounted for the impact of 'crossovers' between IHD and CRRT found no improvement in clinical outcomes with the delivery of higher intensity RRT, including endpoints such as 60-day survival, renal recovery and (in a subsequent publication) health-related quality of life [21].

This convincing result was confirmed by another large multicenter trial of higher versus standard dose RRT in critically ill patients with AKI, restricted to those receiving CRRT (used almost exclusively to provide RRT to patients in ICUs in Australia and New Zealand centers, where the trial was conducted) [22]. The Randomized Evaluation of Normal versus Augmented Level ('RENAL') Replacement Therapy Study enrolled patients with AKI receiving CRRT (all with CVVHDF) in 35 ICUs distributed throughout Australia and New Zealand.

The subjects were randomized to higher intensity therapy (prescribed effluent dose of 40 ml/kg/h) versus a lower intensity therapy (25 ml/kg/h). 1,508 patients were randomized: 721 in the higher intensity group and 743 individuals in the lower intensity group. Mortality at 90 days was identical at 44.7% in both groups. Similarly, renal recovery did not differ between the groups: at 28 days, 14.4 vs. 12.2%, p = 0.31; and 90 days, 6.8 vs. 4.4%, p = 0.14 in the higher intensity versus lower intensity groups, respectively.

Taken together, the negative results of these two large multicenter randomized trials do not suggest any clinical benefit (mortality or renal recovery) with the prescription of higher-dose RRT for AKI. Although it is important to ensure effective delivery of the lower prescribed doses that were found to be adequate in these trials, and it is accepted that specific patients (hypercatabolic patients, for example) may require higher dose RRT to achieve adequate metabolic control, a minimum delivered dose of at least 20 ml/kg/h for CRRT and a single-pool Kt/V of 1.2 for IHD thrice weekly appears adequate for many AKI patients.

Recent meta-analyses including all of these trials are similarly convincingly negative, showing no improvement in survival or renal recovery overall or in patient subsets (septic vs. nonseptic, etc.) with higher doses of RRT [23, 24]. Taken together, the combination of two large negative randomized trials and consistently negative meta-analyses suggests that increasing RRT dose as it is currently understood, prescribed, delivered and quantified is unlikely to further improve outcomes compared to current standard recommendations. However, advances in the technology of membranes and RRT equipment may change this conclusion as new data emerge. For example, current and future studies may identify alternative markers of adequate RRT dose, such as clearance of larger molecules, or biomarkers such as leukocyte response to inflammatory stimuli. Of course, not all aspects of RRT 'dosing' are as evidence-based or carefully managed as the control of azotemia. The control of fluid balance is likely to be a major determinant in critically ill patients supported with RRT [25], yet such management has not been protocolized and studied in randomized trials to establish the optimal approach.

Similarly, although RRT modalities vary in the efficacy of their control of abnormalities such as hypocalcemia, metabolic acidosis and dysnatremias, the approach to the use of RRT modalities in correcting these derangements is guided by very little evidence, despite the significant potential for harm with aggressive RRT in these settings. Furthermore, the potential for higher doses of RRT to remove drugs, such as antibiotics, to a greater degree than expected has been a relatively neglected topic of research and practice, and is another example of an area of deficiency in the current acute RRT literature [26–29]. In this regard, the evident inability of centers to adequately supplement and avoid hypophosphatemia in the high-dose group of the ATN trial is of particular concern [20]; if pharmacotherapy is inadequate for an anticipated RRT-induced electrolyte depletion, it raises significant concerns about the potential for RRT to precipitate subtherapeutic concentrations of many drugs (including critical antimicrobials) that are not routinely dosed with therapeutic drug monitoring and adjustment in clinical practice. A more recent publication further underscored such concerns, demonstrating that RRT-induced hypophosphatemia was associated with a greater incidence of failed weaning from mechanical ventilation and higher tracheostomy rates [30].

In summary, renal support for patients with AKI has improved significantly in recent years, and is increasingly guided by evidence-based consensus. Available evidence suggests that patient outcomes are improved with these therapies, but there is still much work to be done to further optimize renal support for patients with AKI. In clinical practice, the provision of acute RRT that effectively improves clinical outcomes (including optimal safety) requires careful, evidence-based prescription and monitoring of dose, including compulsive supplementation of electrolytes and drugs removed along with uremic toxins. Areas of research that are likely to yield further improvements in the outcomes of patients with AKI requiring RRT include early versus late initiation, strategies for the control of fluid balance in AKI, evidence-based approaches to the dosing of medications and correction of acid-base and electrolyte abnormalities during acute RRT, and the identification of important uremic toxins with adaptation of RRT technology to further improve the outcomes of this devastating form of critical illness.

Acknowledgements

Dr. Koyner is supported by K23 DK081616, and Dr. Udani is supported by 5T32DK007510–23.

References

1 Doshi M, Murray PT: Approach to intradialytic hypotension in ICU patients with ARF. Artif Organs 2003;27:772.

2 Mehta RL: Indications for dialysis in the ICU: renal replacement versus renal support. Blood Purif 2001;19:227–232.

3 Murray PT, Hall JB: Renal replacement therapy for acute renal failure. Am J Respir Crit Care Med 2000;162:777.

4 Conger JD: A controlled evaluation of prophylactic dialysis in post-traumatic acute renal failure. J Trauma 1975;15: 1056–1063.

5 Gillum DM, Dixon BS, Yanover MJ, et al: The role of intensive dialysis in acute renal failure. Clin Nephrol 1986;25:249.

6 Bouman CS, Oudemans-Van Straaten HM, Tijssen JG, et al: Effects of early high-volume continuous venovenous hemofiltration on survival and recovery of renal function in intensive care patients with acute renal failure: a prospective, randomized trial. Crit Care Med 2002;30:2205.

7 Gettings LG, Reynolds HN, Scalea T: Outcome in post-traumatic acute renal failure when continuous renal replacement therapy is applied early vs late. Intensive Care Med 1999;25:805–813.

8 Demirkiliç U, Kuralay E, Yenicesu M, et al: Timing of replacement therapy for acute renal failure after cardiac surgery. J Card Surg 2004;19:17–20.
9 Elahi MM, Lim MY, Joseph RN, et al: Early hemofiltration improves survival in postcardiotomy patients with acute renal failure. Eur J Cardiothorac Surg 2004;26:1027–1031.
10 Wu VC, Wo KJ, Chang HW, et al: Early renal replacement therapy in patients with postoperative acute liver failure associated with acute renal failure: effect on postoperative outcomes. J Am Coll Surg 2007;205:266–276.
11 Liu KD, Himmelfarb J, Paganini E, et al: Timing of initiation of dialysis in critically ill patients with acute kidney injury. Clin J Am Soc Nephrol 2006;1:915–919.
12 Bouchard J, Soroko SB, Chertown GM, et al: Fluid accumulation, survival and recovery of kidney function in critically ill patients with acute kidney injury. Kidney Int 2009;76: 422–427.
13 Evanson JA, Himmelfarb J, Wingard R, et al: Prescribed versus delivered dialysis in acute renal failure patients. Am J Kidney Dis 1998;32:731.
14 Paganini EP, Tapolyai M, Goormastic M, et al: Establishing a dialysis therapy/patient outcome link in intensive care unit acute dialysis for patients with acute renal failure. Am J Kidney Dis 1996;28:S81.
15 Schiffl H, Lang SM, Fischer R: Daily hemodialysis and the outcome of acute renal failure. N Engl J Med 2002;346:305.
16 Ronco C, Bellomo R, Homel P, et al: Effects of different doses in continuous veno-venous haemofiltration on outcomes of acute renal failure: a prospective randomised trial. Lancet 2000;356:26.
17 Tolwani AJ, Campbell RC, Stofan BS, et al: Standard versus high-dose CVVHDF for ICU-related acute renal failure. J Am Soc Nephrol 2008;19:1233–1238.
18 Faulhaber-Walter R, Hafer C, Jahr N, et al: The Hanover Dialysis Outcome study: comparison of standard versus intensified extended dialysis for treatment of patients with acute kidney injury in the intensive care unit. Nephrol Dial Transplant 2009;24:2179–2186.
19 Saudan P, Niederberger M, De Seigneux S, et al: Adding a dialysis dose to continuous hemofiltration increases survival in patients with acute renal failure. Kidney Int 2006;70:1312–1317.
20 VA/NIH Acute Renal Failure Trial Network, Palevsky PM, Zhang JH, O'Connor TZ, et al: Intensity of renal support in critically ill patients with acute kidney injury. N Engl J Med 2008;359:7–20.
21 Johansen KL, Smith MW, Unruh ML, et al: Predictors of health utility among 60-day survivors of acute kidney injury in the Veterans Affairs/National Institutes of Health Acute Renal Failure Trial Network Study. Clin J Am Soc Nephrol 2010;5: 1366–1372.
22 The RENAL Replacement Study Investigators, Bellomo R, Cass A, Cole L, et al: Intensity of continuous renal replacement therapy in critically-ill patients. N Engl J Med 2009;361:1627–1638.
23 Van Wert R, Friedrich JO, Scales DC, et al: High-dose renal replacement therapy for acute kidney injury: Systematic review and meta-analysis. Crit Care Med 2010;38:1360–1369.
24 Jun M, Lambers Heerspink HJ, Ninomiya T, et al: Intensities of renal replacement therapy in acute kidney injury: a systematic review and meta-analysis. Clin J Am Soc Nephrol 2010;5:956–963.
25 Bouchard J, Soroko SB, Chertow GM, et al: Fluid accumulation, survival, and recovery of kidney function in critically ill patients with acute kidney injury. Kidney Int 2009;76: 422–427.
26 Li AM, Gomersall CD, Choi G, Tian Q, Joynt GM, Lipman JJ: A systematic review of antibiotic dosing regimens for septic patients receiving continuous renal replacement therapy: do current studies supply sufficient data? Antimicrob Chemother 2009;64:929–937.
27 Schetz M: Drug dosing in continuous renal replacement therapy: general rules. Curr Opin Crit Care 2007;13:645–651.

28 Bouman CS, van Kan HJ, Koopmans RP, et al: Discrepancies between observed and predicted continuous venovenous hemofiltration removal of antimicrobial agents in critically ill patients and the effects on dosing. Intensive Care Med 2006;32:2013–2019.
29 Soy D, Torres A: Antibacterial dosage in intensive-care-unit patients based on pharmacokinetic/pharmacodynamic principles. Curr Opin Crit Care 2006;12:477–482.
30 Demirjian S, Teo BW, Guzman JA, et al: Hypophosphatemia during continuous hemodialysis is associated with prolonged respiratory failure in patients with acute kidney injury. Nephrol Dial Transplant 2011, E-pub ahead of print.

Patrick T. Murray, MD, FASN, FRCPI, FJFICMI
Prof. of Clinical Pharmacology, University College Dublin
Consultant in Nephrology and Clinical Pharmacology, Mater Misericordiae University Hospital
UCD Catherine McAuley Centre, Nelson Street, Dublin 7 (Ireland)
Tel. +353 1 716 4519, E-Mail patrick.murray@ucd.ie

Kellum JA, Ronco C, Vincent J-L (eds): Controversies in Acute Kidney Injury.
Contrib Nephrol. Basel, Karger, 2011, vol 174, pp 222–231

Why Did Dose Trials Fail?

Michael Joannidis

Medical Intensive Care Unit, Department of Internal Medicine I, Medical University Innsbruck, Innsbruck, Austria

Abstract

The search for the optimal intensity in renal replacement therapy of the critically ill has resulted in several well-designed randomized controlled trials over the last decade. The findings, however, have been somewhat disappointing. Whereas earlier trials indicated a survival benefit using higher doses (>35 ml/kg/h) or daily dialysis, more recent and much larger trials have shown that intensities in the range of 20–25 ml/kg/h or alternate day dialysis may be sufficient. This review attempts to explain these discrepant findings by critically reappraising several issues including the concept of dose, comparability of doses, importance of middle molecule clearance, indication and timing of treatment as well as patient heterogeneity between the published studies.

'Dosis venenum facit' (Paracelsus, 1493–1541)

In medicine the use of any treatment demands an understanding of the dose to be used in a given patient – intensive care medicine is no exception. Additionally, dose-effect relationships often demonstrate a dose above which 'more' means 'harm'. With respect to renal replacement therapy (RRT), dose is considered as the intensity of the respective applied modality. After their introduction into intensive care medicine, both intermittent and continuous renal replacement therapies were characterized by low doses mainly due to technical constraints, although with further development almost any intensity became theoretically possible by the end of the 20th century.

In search of the optimal dose, randomized controlled trials applying higher doses as well as varying treatment frequencies were performed after the millennium, which seemed to suggest an improved survival when doses of up to 45 ml/kg/h in continuous RRT (CRRT) [1, 2] or daily hemodialysis were used [3].

These trials, however, were relatively small single-center studies and confirmation by larger trials was warranted. Two large multicenter trials, the ATN [4] and RENAL studies [5], were published in 2008 and 2009, respectively, accompanied by a smaller investigation [6] – all demonstrating the same findings. These trials indicated that CRRT doses above 25 ml/kg/h or high-intensity intermittent hemodialysis/sustained low efficiency dialysis may not offer any advantage to critically ill patients. The discrepancies between the findings of all these well-designed trials may be interpreted in several ways. Either the earlier trials showed an effect which was found by chance (type I error) or the consecutive large trials might have missed an important signal (type II error). Alternatively, the comparability of the studies could have been hampered by inherent differences in the specific study designs.

Is the Applied Concept of Dose Valid in Critically Patients?

When considering dose in RRT, this usually implies small solute clearance and this is mostly urea clearance. Small solute clearance may be considered equal for diffusive (e.g. continuous venovenous hemodialysis) and convective methods [e.g. continuous venovenous hemofiltration (CVVH)] [7]. Since the sieving coefficient for small solutes approximates to roughly 1, effluent rates are usually taken as the dose delivered ideally corrected for losses of efficacy when predilution is employed. Recent investigations, however, indicate that this procedure may overestimate actual urea clearances in continuous venovenous hemodiafiltration by 24% [8]. The clearance of larger molecules may be significantly influenced by the method used, usually being higher when convective techniques (CVVH) are applied, though recent in vitro studies do question even this concept [9]. Up to now, however, no randomized controlled trial has specifically addressed this problem [10].

A second unresolved issue is that intermittent methods use single pool Kt/V for dose estimations. Standard Kt/V requires that patients are in a steady state with respect to urea. This is not the case in acute kidney injury. Additionally, in critically ill patients the accuracy of Kt/V is compromised by the dissociation between the distribution volume of urea and total body water [11].

Finally, no study has yet validated that clearances provided by single pool Kt/V and effluent rates adjusted for body weight are comparable. For 'semicontinuous' RRT like sustained low efficiency dialysis, a dose concept does not even exist.

Was the Importance of Convection Missed?

Improved survival associated with increased intensity shown by Ronco et al. [1] is frequently attributed to increased middle molecule clearance. This was the only study specifically investigating the effects of CVVH, a purely convective

Table 1. Characteristics of larger randomized controlled trials (n >150) investigating intensity (dose) of RRT

Study	Patients n	Centers	Modality	Prescribed dose ml/kg/h	Delivered dose[1] ml/kg/h	Convective dose ml/kg/h
Ronco et al. [1]	425	1 (2 ICU)	CVVH postdilution	20 35 45	18 32 41	18 32 41
Schiffl et al. [3]	160	1	IHD	weekly Kt/V	3.0 5.8	NA NA
Saudan et al. [2]	206	1	CVVH/ CVVHDF predilution	25 42	17 28	17 15
Tolwani et al. [6]	200	1	CVVHDF predilution	20 35	17 29	7 13
Palevsky et al. [4]	1,124	27	CVVHDF predilution	20 35	19 31	10 16
Bellomo et al. [5]	1,508	35	CVVHDF postdilution	25 40	22 33	12 17
Faulhaber-Walter et al. [13]	157	1 (7 ICU)	SLED	d SLED int SLED	39.2 up to 68.4	?

ICU = Intensive care unit; IHD = intermittent hemodialysis; CVVHDF = continuous venovenous hemodiafiltration; d SLED = daily sustained low efficiency dialysis; int SLED = intensified sustained low efficiency dialysis.

[1] Delivered dose is the prescribed dose corrected for actual delivery (within 24 h) and predilution if applicable.

method comparing prescribed doses of 45 or 35 ml/kg/h versus 20 ml/kg/h. Improved survival was reported for those receiving 35 ml/kg/h as compared to 20 ml/kg/h with no further improvement observed at 45 ml/kg/h. Two facts, however, argue against increased middle molecule clearance as a key factor. First, Saudan et al. [2] demonstrated improvement in survival solely by the addition of a dialysate flow at 18 ml/kg/h to the filtration rate of approximately 25 ml/kg/h used in the control group. Secondly, although the study by Ronco et al. [1] did utilize approximately twice the convective doses administered to the control groups in both the ATN [4] and RENAL [5] trials, mortality in their control group was still 11 and 26% higher as compared to the respective control groups of the ATN and RENAL trials (table 1). Consequently, increasing middle molecule clearance may not confer the presumed benefit in acute kidney injury.

APACHE II/ III (SD)	Mortality %	Initiation (day of ICU admission)	BUN mg/dl (mmol/l)	Creatinine mg/dl (µmol/l)	Sepsis %	Oliguria %
23 (3)	59 (30d) 42 43	NA	50 (18)	3.6 (318)	12	100
87 (III)	46 28	NA	90 (32)	5.0 (442)	36	46
25 (9)	61 (90d) 46	NA	83 (30)	4.8 (428)	60	37
26 (7)	60 64	8d	76 (27)	4.3 (376)	54	64
26 (7.2)	51.5 (60d) 53.6	6–7d	66 (24)	4.1 (362)	63	78
102 (III) (26)	44.7 (90d) 44.7	2.1d	64 (23)	3.8 (334)	49	60
32 (7.6)	70.7 70.4	6.2 -5.5d	63 (22)	3.1 (273)	72	73

Did the Studies Investigate Equivalent Doses?

The consecutive CRRT studies tried to investigate comparable dose ranges, prescribing 20–25 ml/kg/h in the control arm and 35–45 ml/kg/h in the intervention groups [1, 2, 4–6]. However, due to differences in dose delivery as well as variance in applying pre- and postdilutional modes, the effective doses achieved demonstrated considerable variations of 17–22 and 27–42 ml/kg/h in the control and intensive treatment groups, respectively (table 1). The ATN [4] trial also allowed application of both intermittent and continuous modalities, which are hardly comparable as discussed above.

Delivered doses below 20 ml/kg/h should be considered problematic for severely ill patients. Brause et al. [12] could demonstrate by exact determination of delivered Kt/V during CVVH that a daily Kt/V of 0.56 (corresponding to a dose of 13 ml/kg/h) was insufficient to control the acid-base balance and azotemia, whereas a higher Kt/V of 0.8 (corresponding to 20 ml/kg/h) established homeostasis even in the sicker patients. The beneficial effects of increased dose in the studies by Ronco et al. [1] and Saudan et al. [2] were observed in patient cohorts with very high mortality rates of 63% (30 days) and 61% (90 days) in the control groups, which is considerably higher than those observed in the larger trials where mortalities of 51.5% (60 days, ATN) and 44.7% (90 days, RENAL) were reported despite a higher severity of disease according to the respective APACHE scores and rates of sepsis [4, 5]. Taking into account an approximate 90% delivery of prescribed dose [1, 2] as well as the effects of predilution [2], roughly 18 and 17 ml/kg/h were delivered in the control groups of these two CRRT trials and showed benefit by higher intensities (table 1). Consequently, undertreatment of control groups must be taken into consideration. The same might be true for the study comparing daily to alternate day hemodialysis [3] in which patients in the control group received a Kt/V of only 0.92 per session as compared to a Kt/V of 1.3 in the ATN trial.

Should a Fixed Dose Be Applied without Any Feedback from a Target Parameter?

A completely different question is whether the methodology used in all these trials employed a correct approach [1, 2, 4–6]. All the studies used fixed prescribed doses independent from acid-base status, volume status and level of uremia. In medicine the dose employed in most treatments is usually tailored to a response of a therapeutic target. For example, a target mean arterial pressure is aimed for when administering a vasopressor, dosing of antibiotics is adjusted to blood levels, etc. Interestingly, this concept was not applied in the large dose studies. Only one recent study, the Hannover Outcome trial, tailored dose to a certain blood urea nitrogen (BUN) target [13]. Interestingly, they could not find any difference in survival when BUN <45 or 60–75 mg/dl was targeted, achieved by increasing intensities of sustained low efficiency dialysis. The results of this study are somewhat hampered by a relatively low number of patients (156 patients) with an extremely high mortality of roughly 70%. Nevertheless, a significant signal does not appear to be present and tailoring dose to a certain BUN level may not really be necessary. Of note, both the ATN and the RENAL trials achieved BUN levels below 47 mg/dl even in the lower intensity treatment groups [4, 5].

It must, however, be considered that BUN may not be the right target parameter. Urea itself is relatively nontoxic, with its concentration dependent on volume of distribution, production rate and tubular reabsorption. The volume of distribution for urea increases significantly in the volume-overloaded critically ill patient [14] leading to lower serum concentrations, although this may not be the same for other uremic toxins not routinely measured. Also, increased BUN has been shown to be a 'biomarker' for increased mortality in several situations including acute heart failure, coronary artery disease, stroke, severe pneumonia, bone marrow transplantation, acute pancreatitis and following esophagectomy. Therefore, serum levels of urea may not necessarily reflect the concentrations of uremic toxins and therefore may by considered a poor marker for uremia treatment [15].

None of the studies on dose of RRT, however, investigated tailoring a delivered dose. It could be envisaged that application of an initial 'high dose' followed by a 'standard dose' may decrease the time period during which homeostasis is severely disturbed.

Was There a Significant Heterogeneity between the Study Populations?

The dose [1, 2, 4–6] studies investigated patients with different disease severity with APACHE II scores ranging from 23–26 and APACHE III scores ranging from 84–102. Also, the rate of sepsis varied from 12 to 63% of patients (table 1). Interestingly, the study which showed the largest benefit from increasing doses had the lowest number of septic patients (i.e. 15%) and the lowest APACHE II score (i.e. 23) with the smallest standard deviation but the highest mortality in the control group [1].

Finally pre-existing chronic kidney disease was included in some studies [1, 5] and excluded in the remaining studies [2–4, 6]. Patients with chronic kidney disease are known to show lower mortality and higher rates of nonrecovery. Despite this, it could be demonstrated by subgroup analysis that there was no difference in outcome between patients with a glomerular filtration rate <60 or >60 ml/min [5].

Although heterogeneity definitely remains to be an issue, a recent meta-analysis could not find any benefit of high intensity for any subgroup addressed [16].

Was the Influence of Timing on Treatment Acknowledged Sufficiently?

Timing of treatment should not be considered in isolation but together with the treatment dose applied. This, in turn, will determine the time needed until the control of uremic waste products is achieved.

The effect of applied dose might be greatest in those with the most severe derangement of homeostasis. The ATN trial did not even allow such a severe disturbance to occur because prestudy RRT was allowed if required and was performed in 65% of the patients. With regard to the average BUN or creatinine at initiation, all dose trials [1–6] started RRT with BUN between 50 and 90 mg/

dl and serum creatinine between 3.4 and 5 mg/dl, which may be regarded quite timely [17].

Large differences between the studies were found in initiation, which was defined as day after intensive care unit admission ranging from 2 [5] to 8 days [6]. This parameter, however, is misleading when used with the intention to define timing. Approximately 20% of patients develop acute kidney injury with the need for RRT late in their admission, which is associated with inferior outcomes [18, 19]. Furthermore, this criterion (i.e. duration of admission to intensive care unit to start of RRT) can only be determined retrospectively and cannot be applied prospectively to guide treatment.

With regard to duration and termination of RRT, only the large dose trials (ATN and RENAL) predefined specific termination procedures. The ATN trial continued RRT as long as the creatinine clearance was below 12 ml/min and discontinued if creatinine clearance was greater than 20 ml/min. However, the duration of RRT therapy varied significantly from 6 to 13 days between studies.

In summary, all the reported trials appeared to initiate RRT timely, which may have reduced effects on the importance of applied dose.

Could the Indications for Renal Replacement Therapy Have Influenced Outcome?

Currently no single standardized indication for RRT can be given. Commencement of RRT usually depends on features of homeostatic disturbance including uremia, azotemia, acidosis and electrolyte disturbance or oligo-/anuria. Whereas metabolic derangements may respond to dose, oliguria and its adverse effects on volume overload may require further fluid rather than solute removal. Thus, it is interesting to note that the rate of oliguria was very high in trials, ranging from 37% [2] to 100% [1] of the patients. The RENAL trial, for example, applied oliguria and severe edema in 60 and 44%, respectively, as criteria to initiate RRT [5]. It could be hypothesized that the influence of dose may be irrelevant in patients requiring fluid removal. Of note, only 37% of the patients were oliguric in the study by Saudan et al. [2], indicating a predominantly metabolic disturbance in this patient group who in turn may have benefited from an improved outcome from higher dose.

Could Significant Adverse Effects of High Dose Continuous Renal Replacement Therapy Have Impaired Outcome?

Application of RRT is associated with adverse effects due to the removal of nutrients, especially amino acids, trace elements, vitamins and electrolytes [20, 21]. Hypophosphatemia was the most frequent adverse event reported in the intensive strategies groups of both the ATN trial, amounting to 17% (vs. 10% in the less intensive group), and the RENAL trial with 65% in the higher intensity group versus 54% in the control group. Hypophosphatemia may result in decreased cardiac output [22] and (respiratory) muscle weakness [23]. It is therefore regarded as a serious disturbance in the critically ill and is even reported to be associated with worse outcome [24], especially in patients with sepsis for whom an 8-fold increased risk of death was reported [25]. Furthermore, removal of antibiotics is increased when applying higher-dose RRT resulting in lower plasma and, hence, therapeutic levels. It may be speculated that septic patients may be more susceptible to adverse effects of high-intensity CRRT, i.e. excessive removal of antibiotics, vitamins and trace elements [20, 21], and more significant impact on hemodynamic stability. A recent randomized controlled multicenter trial included 80 nonoliguric patients with severe sepsis or septic shock showed that early initiation of CVVH at a dose of 25 ml/kg/h may be deleterious as compared to standard medical treatment [26].

Altogether, it may be construed that adverse effects associated with sustained higher intensity treatment may have outperformed beneficial effects to some degree.

Do Multicenter Trials Miss a Significant Signal?

The intensive care literature is littered with examples of significant findings demonstrated in single-center studies which could not be reproduced at a multicenter level. Perhaps the most quoted is that of intensive insulin therapy which significantly reduced mortality in large single-center trials [27, 28], but failed when applied in a multicenter approach [29]. The significance of any demonstrated differences between two means/medians depends on the variation within the whole sample. Single centers may do an excellent job applying a specific intervention in a quite homogenous setting characterized by a specific patient selection as well as standardized treatment modalities for these. Including several centers in a study increases noise to signal ratio by increasing patient heterogeneity in terms of case mix, severity of disease and efficiency of treatment modalities. Unfortunately, mortality as an endpoint does not provide a measure for variation.

This may be partly overcome by decreasing the chance of a type II error. The VA/ATN trial had a power of 90% to detect a 10% difference in survival and the RENAL trial had a 90% power to detect a 8.5% absolute decrease in 90-day mortality, which should suffice to detect the 16–20% decrease in mortality reported in the two single-center trials [1, 2]. Despite this, there is roughly a 5% chance of missing an existing improvement in mortality by increasing treatment dose. Of note a recent meta-analysis on dose trials only just missed demonstrating a benefit of high-intensity treatment modalities with a RR of 0.89 (CI: 0.77–1.03) [16].

Conclusion

The dose trials only failed in that they did not support the hypothesis that more is necessarily better. The discrepancies between the findings of published dose trials may be explained by the limitations of the current concept of dose, differences in dose delivery, timing, indication, and patient case mix, or relevant influence of adverse effects of high-dose RRT. They have succeeded by providing the important information that delivered doses lower than 20–22 ml/kg/h may be harmful to patients, whereas a sustained increased dose over the whole period of renal failure may not be helpful in improving outcome. It became apparent that higher doses are associated with higher rates of adverse effects.

It is still feasible that specific patients may require higher doses for a certain time period.

References

1 Ronco C, Bellomo R, Homel P, Brendolan A, Dan M, Piccinni P, La Greca G: Effects of different doses in continuous veno-venous haemofiltration on outcomes of acute renal failure: a prospective randomised trial. Lancet 2000;356:26–30.

2 Saudan P, Niederberger M, De SS, Romand J, Pugin J, Perneger T, Martin PY: Adding a dialysis dose to continuous hemofiltration increases survival in patients with acute renal failure. Kidney Int 2006;70:1312–1317.

3 Schiffl H, Lang SM, Fischer R: Daily hemodialysis and the outcome of acute renal failure. N Engl J Med 2002;346:305–310.

4 Palevsky PM, Zhang JH, O'Connor TZ, Chertow GM, Crowley ST, Choudhury D, Finkel K, Kellum JA, Paganini E, Schein RM, Smith MW, Swanson KM, Thompson BT, Vijayan A, Watnick S, Star RA, Peduzzi P: Intensity of renal support in critically ill patients with acute kidney injury. N Engl J Med 2008;359:7–20.

5 Bellomo R, Cass A, Cole L, Finfer S, Gallagher M, Lo S, McArthur C, McGuinness S, Myburgh J, Norton R, Scheinkestel C, Su S: Intensity of continuous renal-replacement therapy in critically ill patients. N Engl J Med 2009;361:1627–1638.

6 Tolwani AJ, Campbell RC, Stofan BS, Lai KR, Oster RA, Wille KM: Standard versus high-dose CVVHDF for ICU-related acute renal failure. J Am Soc Nephrol 2008;19:1233–1238.

7 Ricci Z, Ronco C, Bachetoni A, D'Amico G, Rossi S, Alessandri E, Rocco M, Pietropaoli P: Solute removal during continuous renal replacement therapy in critically ill patients: convection versus diffusion. Crit Care 2006;10:R67.

8 Granado RC, Macedo E, Chertow GM, Soroko S, Himmelfarb J, Ikizler TA, Paganini EP, Mehta RL: Effluent volume in continuous renal replacement therapy overestimates the delivered dose of dialysis. Clin J Am Soc Nephrol 2011;6:467–475.

9 Messer J, Mulcahy B, Fissell WH: Middle-molecule clearance in CRRT: in vitro convection, diffusion and dialyzer area. ASAIO J 2009;55:224–226.

10 Pannu N, Klarenbach S, Wiebe N, Manns B, Tonelli M: Renal replacement therapy in patients with acute renal failure: a systematic review. JAMA 2008;299:793–805.

11 Ikizler TA, Sezer MT, Flakoll PJ, Hariachar S, Kanagasundaram NS, Gritter N, Knights S, Shyr Y, Paganini E, Hakim RM, Himmelfarb J: Urea space and total body water measurements by stable isotopes in patients with acute renal failure. Kidney Int 2004;65:725–732.

12 Brause M, Neumann A, Schumacher T, Grabensee B, Heering P: Effect of filtration volume of continuous venovenous hemofiltration in the treatment of patients with acute renal failure in intensive care units. Crit Care Med 2003;31:841–846.

13 Faulhaber-Walter R, Hafer C, Jahr N, Vahlbruch J, Hoy L, Haller H, Fliser D, Kielstein JT: The Hannover Dialysis Outcome study: comparison of standard versus intensified extended dialysis for treatment of patients with acute kidney injury in the intensive care unit. Nephrol Dial Transplant 2009;24:2179–2186.
14 Macedo E, Bouchard J, Soroko SH, Chertow GM, Himmelfarb J, Ikizler TA, Paganini EP, Mehta RL: Fluid accumulation, recognition and staging of acute kidney injury in critically-ill patients. Crit Care 2010;14:R82.
15 Joannidis M, Forni LG: Clinical review: timing of renal replacement therapy. Crit Care 2011;15:233.
16 Berger MM, Shenkin A, Revelly JP, Roberts E, Cayeux MC, Baines M, Chiolero RL: Copper, selenium, zinc, and thiamine balances during continuous venovenous hemodiafiltration in critically ill patients. Am J Clin Nutr 2004;80:410–416.
17 Druml W: Nutritional management of acute renal failure. J Ren Nutr 2005;15:63–70.
18 Payen D, Mateo J, Cavaillon JM, Fraisse F, Floriot C, Vicaut E: Impact of continuous venovenous hemofiltration on organ failure during the early phase of severe sepsis: a randomized controlled trial. Crit Care Med 2009;37:803–810.
19 Van Wert R, Friedrich JO, Scales DC, Wald R, Adhikari NK: High-dose renal replacement therapy for acute kidney injury: Systematic review and meta-analysis. Crit Care Med 2010;38:1360–1369.
20 Liu KD, Himmelfarb J, Paganini E, Ikizler TA, Soroko SH, Mehta RL, Chertow GM: Timing of initiation of dialysis in critically ill patients with acute kidney injury. Clin J Am Soc Nephrol 2006;1:915–919.
21 Bagshaw SM, Uchino S, Bellomo R, Morimatsu H, Morgera S, Schetz M, Tan I, Bouman C, Macedo E, Gibney N, Tolwani A, Oudemans-van Straaten HM, Ronco C, Kellum JA: Timing of renal replacement therapy and clinical outcomes in critically ill patients with severe acute kidney injury. J Crit Care 2009;24:129–140.
22 Joannidis M, Metnitz PG: Epidemiology and natural history of acute renal failure in the ICU. Crit Care Clin 2005;21:239–249.
23 Zazzo JF, Troche G, Ruel P, Maintenant J: High incidence of hypophosphatemia in surgical intensive care patients: efficacy of phosphorus therapy on myocardial function. Intensive Care Med 1995;21:826–831.
24 Gravelyn TR, Brophy N, Siegert C, Peters-Golden M: Hypophosphatemia-associated respiratory muscle weakness in a general inpatient population. Am J Med 1988;84:870–876.
25 Geerse DA, Bindels AJ, Kuiper MA, Roos AN, Spronk PE, Schultz MJ: Treatment of hypophosphatemia in the intensive care unit: a review. Crit Care 2010;14:R147.
26 Shor R, Halabe A, Rishver S, Tilis Y, Matas Z, Fux A, Boaz M, Weinstein J: Severe hypophosphatemia in sepsis as a mortality predictor. Ann Clin Lab Sci 2006;36:67–72.
27 Van den Berghe G, Wilmer A, Hermans G, Meersseman W, Wouters PJ, Milants I, Van WE, Bobbaers H, Bouillon R: Intensive insulin therapy in the medical ICU. N Engl J Med 2006;354:449–461.
28 Van den Berghe G, Wouters P, Weekers F, Verwaest C, Bruyninckx F, Schetz M, Vlasselaers D, Ferdinande P, Lauwers P, Bouillon R: Intensive insulin therapy in the critically ill patients. N Engl J Med 2001;345:1359–1367.
29 Finfer S, Chittock DR, Su SY, Blair D, Foster D, Dhingra V, Bellomo R, Cook D, Dodek P, Henderson WR, Hebert PC, Heritier S, Heyland DK, McArthur C, McDonald E, Mitchell I, Myburgh JA, Norton R, Potter J, Robinson BG, Ronco JJ: Intensive versus conventional glucose control in critically ill patients. N Engl J Med 2009;360:1283–1297.

Prof. Michael Joannidis, MD
Medical Intensive Care Unit, Department of Internal Medicine I
Medical University Innsbruck
Anichstrasse 35, 6020 Innsbruck (Austria)
Tel. +43 512 504 24181, E-Mail MICHAEL.JOANNIDIS@I-MED.AC.AT

Kellum JA, Ronco C, Vincent J-L (eds): Controversies in Acute Kidney Injury.
Contrib Nephrol. Basel, Karger, 2011, vol 174, pp 242–251

Renal Replacement Therapy for Acute Renal Injury: We Need Better Therapy

Savag G. Demirjian · Emil P. Paganini

Critical Care Nephrology, Department of Nephrology/Hypertension, Glickman Urological and Kidney Institute, The Cleveland Clinic Foundation, Cleveland, Ohio, USA

Abstract

Dialytic support of patients with acute kidney injury (AKI) has taken on an important aspect of critical care medicine. Increased morbidity and mortality associated with the AKI syndrome and the lack of great improvement despite the addition of differing dialytic techniques (and intensity) speaks to the need for a re-evaluation of renal support. Continuous therapies have brought greater control of urea, volume, acid/base status and enhanced hemodynamic stability over the traditional intermittent approaches. However, the incremental efficiency achieved by intense dialysis has not improved outcome in patients with AKI. We need to move beyond urea-based decision-making and pursue clinically relevant goal-targeted therapies. The latter will invariably lead to re-evaluation of the timing, intensity and duration of therapy, which traditionally have been mainly solute driven. Whether this will be via specifically designed membrane extracorporeal support or focused drug or cell-based therapies is currently under consideration. Volume determination and variability remain another moving target for therapy. Machine-generated feedback mechanisms responding to specific endpoints or compartmental changes are also under development. Improved diagnostic criteria, especially in septic-induced renal dysfunction, may allow for specific adsorption techniques using a variety of membrane-imbedded substances from activated charcoal to polymyxin B to newer resins. Cascade apheretic techniques have been attempted in specific disease entities to capture a larger group of potential toxins, while nanoporous membranes have been developed to remove a specific sized entity. Bio-artificial systems utilizing functioning cells should help with the recovery of injured cell and cell protection in those yet viable. Simple maneuvers to reduce the cost of delivered therapy, and the development of a more robust severity scoring system to help address the futile use of technology would be of great help. Greater attention to elements lost during intervention which may require supplementation, as well as the development of on-line replacement technology and coagulation friendly systems, will help eliminate much of the current cost of therapy.

An exciting beginning to renal support was the use of extracorporeal therapy for the treatment of acute renal 'failure' in the late 1940s and early 1950s. Conceptually, the control of high levels of urea should be associated with an improved status and thus allow for the natural 'healing process' to occur. It was not until renal transplantation was experimentally attempted and the development of a viable long-term blood access that the use of 'long-term' dialytic maneuvers were shown to be helpful in maintaining life in patients awaiting transplantation. Both areas were satisfied with a simple bridge to either re-establishing renal function [acute kidney injury (AKI)] or an attempt at transplant [end-stage renal disease (ESRD)]. Technology was inadequate for both long-term access for chronic support and for lower extracorporeal blood volume for the acute. Attempts at offering 'continuous dialysis' was reported by Scribner in the 1950s [1], but the material and technology were too cumbersome for any effective mode of delivery.

A reliance on intermittent methods of dialysis was used in the early 1960s and 1970s. The concept of dose delivery began with a mass transfer dosage of therapy still using urea as the marker substance [2]. This gave way to a clearance methodology again of urea, known as Kt/V_{urea}, based on the National Cooperative Dialysis Study of the late 1970s and early 1980s [3, 4]. The discussion of whether a 'peak hypothesis' has merit over a 'time-averaged' approach is still debated among clinicians [5]. While Kt/V_{urea} was perhaps the most useful development in the delivery of dialysis, as it allowed for comparative dosing studies and related outcome to dose comparisons, it has kept dialysis development attached to the concept of urea clearance as the surrogate for newer dialytic techniques. In a number of recent prospective randomized studies in both the acute [6, 7] and chronic [8] arenas, increasing urea clearance alone has not been accompanied with improved outcome once a baseline level of support has been obtained. Thus, the search for other indicators of dialysis effect which may be associated with improved outcome has been undertaken.

Technology Advances in the Past

Machine and membrane-based changes have been the focus of system development. Moving from a variable volume blood side and a large rotating drum with semi-single pass dialysate bathing to a more ridged plate design and finally to a capillary system with a single pass dialysate flow system held advantages for both efficiency and patient safety. Membrane development and altered construction, along with a focus on sieving characteristics and biocompatibility again led to a variety of altered endpoints beyond urea control. The addition of volumetric controls added to the precision of the machines and thus allowed for a more stable dialysis experience.

Table 1. Potential advances within reach today

1	Access development with the ability for high blood flow and reduced infection
2	System biocompatibility to reduce clotting and complement activation
3	On-line dialysate/hemofiltration replacement solution with the ability to alter composition
4	Electronic medical record connection to allow for data capturing beyond machine characteristics
5	Enhanced volume control to avoid machine- or practice-generated calculation errors

What followed was a number of technical adjustments (variable dialysate sodium concentration, variable ultrafiltration methods, adjustment of temperature) aimed at patient stability. Membrane variants (high-flux, adsorption-imbedded material, membrane porosity changes to allow for apheresis techniques) targeting a variety of presumed toxin removal became the mantra in both acute and chronic dialytic efforts. Clearance data based on substances other than urea, and the rate of toxin removal following a specific end-treatment goal became practice variations. The introduction of pure convection or combination diffusion/convection to better accommodate a broader array of removal substances was adapted in both the intermittent and continuous therapy sessions. Beyond the intermittent dialytic approach where combination therapy has shown some advantage, no such advantage has been noted in the continuous format of delivery.

Intermittent dialysis machinery became very complex and required the help of specially trained personnel to operate them. Although originally established as an arteriovenous format, the introduction of machine-based venovenous continuous forms of therapy allowed avoiding the entry into major arterial accesses, but brought a bit more complexity to the table. The addition of a variety of possible therapy variants in the same machine added to the complexity of the bedside sessions. However, there have been no real data showing an advantage of the more complex over the less complex support methods.

Targeted Technology Aspirations for the Future

Many of the potential changes which would add to both patient safety and improved therapy outcome are possible with today's technology (table 1), while others will require integrated formats and cooperative analysis (table 2).

Timing Related

Biomarker-Adjusted Timing for Initiation. There have been a number of recent articles showing the potential benefit of an 'early' start on renal support in AKI

Table 2. Potential advances in the future

1	Integrated monitor/machine decision-making
2	Incorporation of cell-implanted membranes for organ crosstalk activities
3	Early focused therapy specific to individual need rather than generic activity
4	Extension of microelectromechanical system-fabricated nanotechnology for efficient, focused biocompatible support

[9, 10]. While many are based on a specific level of a substance (urea, creatinine) or the rate of change of the substance, it has been universally accepted that these markers are themselves problematic. The emergence of such early markers as NGAL or IL-18 in both urine and serum hold the promise for an earlier indication of renal dysfunction beyond the renal physiological alterations considered 'renal success' [11–13]. Whether a combination of markers will be able to precisely identify both the area of dysfunction as well as the size of the damage is currently under investigation. The predictive value of these markers will hold the key to potential long-term support and thus an approach which favors protection of viable sites for long-term access (and the need for ESRD counseling), or merely the short-term bridge to renal recovery, albeit for several weeks of duration. Decisions on initiation or continuation of support will also need to be considered and perhaps early indications of the potential for recovery may be helpful in avoiding valueless aggressive support techniques.

Support System Alteration for Early Targets. Since there is no real understanding of exactly what we are doing with the current forms of renal support beyond volume, acid/base balance, electrolyte control and urea removal, it may be difficult to establish a specific focus for early dialytic support. When introducing a source of morbidity with any invasive format, we need to be mindful of the negatives of extracorporeal systems associated with AKI. Indeed, there is a doubling of mortality in postcardiac surgery patients with AKI who require dialysis compared to those who do not [14]. Thus, avoiding any form of invasive dialytic sessions may be advantageous. A more likely approach might be halting or reversing deteriorating cellular activity. For example, knowing that cells are in an active phase of apoptosis, the addition of a siRNA specific for P-53 pathway disruption might prevent further progression of the process [15]. If an etiology is established early, beyond avoiding the continuation of the insult, perhaps a targeted session with a specific membrane or technique might avoid the further deterioration into full support need. Promise with early endotoxin adsorption leading to an attenuated septic condition may be the type of 'early' support that should be offered, rather than the application of the entire dialytic armamentarium.

Volume-Related System Development

Volume excess and imbalance has been one of the most frequently associated issues in AKI. Participation in both the etiology of AKI (basis for prerenal conditions) as well as the result of renal dysfunction, volume imbalance has recently been a subject of intense study. Indeed, in patients with acute respiratory distress syndrome, those patients who had a tighter control on volume management did better overall having a reduced mortality [16]. The association of volume excess and outcome is not new to critical care medicine. It is not surprising that technology has focused on volume as a specific target in AKI as well as in other disorders. The challenge is determining the extent of volume excess, and the effect of its removal. A variety of relationships have been established between absolute loss, rate of loss and the variability of intravascular volume. While continuous hematocrit measurements have been the most frequent surrogate for relative intravascular volume changes, other blood determination such as serum albumin or b-type natriuretic peptide have also been used. Transthoracic or transesophageal echocardiographic status, central pressure monitoring (CVP, PAC), pulse contour analysis, and other devices and maneuvers have all played a vital role in establishing volume status and change.

Specific machinery has been developed for the sole purpose of volume removal. While the system is attractive for its simplicity, it lacks the sophistication of total integration. It also requires a very porous membrane and thus will lose larger substances along with the volume. Closed-loop monitoring of the relative and absolute volume status and the rate and amount of ultrafiltration would be a valuable adjunct to the current systems. Indeed, several systems have been tied to changes in hematocrit while others have followed a central pressure change. However, these have not been closed stand-alone systems, but require the interpretation and intervention of the caregiver. Before moving to a more continuous method of volume control, such as an implantable device for ultrafiltration, a better understanding of volume status and requirements is needed. The ultimate goals for volume balance are normalization of organ perfusion, organ turgor, fluid compartmentalization and fluid composition. Related complications include intravascular depletion, altered volume composition and unadjusted medication effects.

Toxin-Related Systems

Urea-Based Indications for Support. Urea-based therapy has given us a method for judging 'adequacy' in what is described as extracorporeal renal support. As noted earlier, urea monitoring has driven the delivery systems to better mimic the depurative function of the 'natural kidney', with membrane cutoffs to mirror those of the glomerulus. This nonselective membrane loss has also allowed substances which may be helpful in renal recovery to be removed. Both essential and nonessential amino acids and a number of water soluble vitamins are examples of recent concern [17]. While supplemental therapy during the dialytic session may help, there may well be other unknown factors which are being removed and thus altering the recovery efforts of the kidney. The urea-based dialysis dose relationship to patient mortality had been the mainstay of delivered dosage [5], but has recently been questioned [6, 7]. Indeed, it was hoped that higher rates of urea clearance and removal would have a potential benefit, but these have failed to be associated with improved patient outcome. This is true for the diffusive, convective and combination therapy formats considered in prospective well-populated studies. In fact, there has never been a convincing study to demonstrate the superiority of one form of extracorporeal delivery to another. The arbitrary dosage floor in AKI uses minimally delivered urea-based clearance dosing from the ESRD population ($Kt/V_{urea} \geq 1.2$). However, the ESRD dose ranges have been established over years of therapy delivery and not over the usual several week-long therapy requirement in AKI. Thus, while practiced as a minimal dose delivery, no prospective evidence has been utilized to support this practice.

Endotoxin Indications for Support. An exciting area of ongoing research continues to be specific membrane adsorption techniques for targeted substances. Perhaps the most developed is the use of immobilized polymyxin B either as an additive to a membrane or as a perfusion element in extracorporeal perfusion [18, 19]. Other elements used over the years but never fully developed include activated charcoal (alone as a perfusate or embedded into a membrane structure), plastic dialysis membrane material itself for hemoperfusion [20, 21] or recently developed resins [22, 23]. Earlier activity with single substance-specific anticytokine medication had positive animal results, but fell short in human testing. While the model has come under question, the cascade of events following an endogenous stimulus has brought into question the potential of a single-element adsorption technique. Perhaps removal of a series of active cytokines might hold more promise than focusing on any single factor. Cascade apheresis has been proposed as another form of extracorporeal intervention having a much larger catchment of substances deemed to be involved in toxic-induced AKI. A variety of differing techniques from simple plasmapheresis to selective cascade membrane pheresis are under study [24].

Perhaps the most innovative techniques to be attempted are the bio-artificial systems which have been tested over the past several years. While the thought behind most of the extracorporeal techniques have been to *remove* toxic material, this therapy extends the idea to also *add* something to the process. The membrane addition of cultured proximal tubular cells, immortalized and yet still biologically active remains a technique which has met with mixed reviews. Initial success in both phase I and early phase II studies [25] were followed with less than convincing data in a prospective randomized phase II evaluation [26]. However, with the recognition of organ-to-organ crosstalk, this approach perhaps applied early in the process of AKI might not only uncover valuable information of cellular repair but promote an attenuated AKI process both in severity as well as length of dysfunction.

Nanotechnology has shown some promise in the development of silicone-based uniform nanopore-sized membranes fabricated using microelectromechanical systems [27]. While the membranes can be manufactured to address any specific pore size, the lack of agreement on which 'toxin' to be removed has continued development of a urea-based model. The biocompatibility of these new membrane types has been shown to be alterable with minimal surface treatment and thus allow for the potential implantability of the system. While this is focused on the chronic dialysis population, the advances in surface activity and newer materials may well lead to application in extracorporeal systems with specific targets. Material manipulation can also help in developing systems with lower clotting propensity than the current plastics used in both bloodlines and membranes [28].

Cost Considerations and Impact on Patient Outcome

One of the most difficult aspects of any critical care situation is identifying those patients who might benefit from aggressive support while recognizing that many patients may not improve despite ongoing invasive support. Attempts at defining a severity index within specific intensive care unit populations have surfaced in a variety of formats and have helped to categorize patient types using population-based analysis. From single-center algorithms to study population-based analysis of indicators associated with a poor outcome, there has not yet been a universally acceptable method or system for identifying those patients who would not benefit [29]. The uniqueness of a specific center or patient makes individual predictions far from acceptable for universal application. Thus AKI support, associated with such comorbidities as end-stage liver failure, severe cerebral injury or advanced age continues to be applied. This has led to lengthy and costly intensive care unit support in circumstances which are ultimately terminal.

Regardless of the potential for advances in AKI focused targets, the extracorporeal systems in use today can be greatly enhanced by several simple considerations. One of the most expensive aspects of any continuous dialysis system is fluid replacement/dialysate. Intermittent systems have been able to generate ultrapure dialysate from relatively inexpensive concentrate. While exposure to this fluid is limited to shorter time frames and separated by a diffusive membrane, the concept of membrane sterilization is quite active in the generation of the dialysate. The addition of redundant filtration cascades should allow for the on-line generation of both dialysate and replacement solution and would greatly reduce the cost of therapy.

Another aspect is system failure. This is primarily due to clotting within the system itself. Thought to be due to turbulent flow, bio-incompatibility and inadequate blood access, the use of agents such as heparin and argatroban, or protocols for citrate anti-coagulation, have had some success in reducing clotting and prolonging system life. The addition of drug-related complications, however, will add to the burden of both the bedside personnel as well as the cost of service. Thus, a system with self-cleaning potential, using nothing more than normal saline or dialysate flushes added to a system with more laminar-flow characteristics are potential alterations that are well within our grasp today.

Finally, gathering real-time data from the support session as well as the integrated monitoring and bedside activity would allow for automated generation of variables which might be helpful in establishing benchmarks for therapy. Access to the electronic medical record offered by the dialytic system should alleviate bedside personnel of some of the tasks associated with support, and allow for a reduced number of attendants who would then be able to focus attention on the patient rather than the devices. To date, we have no universally accepted set of benchmarks for the practice of AKI support.

References

1 Hickman RO, Scribner BH: Application of the pumpless hemodialysis system to infants and children. Trans Am Soc Artif Intern Organs 1962;8:309–314.

2 Scribner BH: Role of the artificial kidney in management of acute renal failure. Northwest Med 1959;58: 555–559.

3 Gotch FA, Sargent JA, Keen ML, Lee M: Individualized, quantified dialysis therapy of uremia. Proc Clin Dial Transplant Forum 1974;4:27–35.

4 Sargent J, Gotch F, Borah M, Piercy L, Spinozzi N, Schoenfeld P, Humphreys M: Urea Kinetics: a guide to nutritional management of renal failure. Am J Clin Nutr 1978;31:1696–702.

5 Levine J, Bernard DB: The role of urea kinetic modeling, TAC_{urea}, and Kt/V in achieving optimal dialysis: a critical reappraisal. Am J Kidney Dis 1990;15: 285–301.

6 VA/NIH Acute Renal Failure Trial Network, Palevsky PM, Zhang JH, O'Connor TZ, Chertow GM, Crowley ST, Choudhury D, Finkel K, Kellum JA, Paganini E, Schein RM, Smith MW, Thompson BT, Vijayan A, Watnick S, Star RA, Peduzzi P: Intensity of renal support in critically ill patients with acute kidney injury. N Engl J Med 2008;359:7–20.

7 RENAL Replacement Therapy Study Investigators, Bellomo R, Cass A, Cole L, Finfer S, Gallagher M, Lo S, McArthur C, McGuinness S, Myburgh J, Norton R, Scheinkestel C, Su S: Intensity of continuous renal replacement therapy in critically ill patients. N Engl J Med 2009;361:1627–1638.

8 Unruh M, Benz R, Greene T, Yan G, Beddhu S, DeVita M, Dwyer JT, Kimmel PL, Kusek JW, Martin A, Rehm-McGillicuddy J, Teehan BP, Meyer KB, Hemo Study Group: Effects of hemodialysis dose and membrane flux on health-related quality of life in the HEMO study. Kidney Int 2004;66:355–366.

9 Macedo E, Mehta RL: Early vs late start of dialysis: it's all about timing. Crit Care 2010;14:112.

10 Gibney RT, Bagshaw SM, Kutsogiannis DJ, Johnson C: When should renal replacement therapy for acute kidney injury be initiated and discontinued? Blood Purif 2008;26:473–484.

11 Devarajan P: Review: neutrophil gelatinase-associated lipocalin: a troponin-like biomarker for human acute kidney injury. Nephrology 2010;15:419–428.

12 Tuladhar SM, Puntmann VO, Soni M, Punjabi PP, Bogle RG: Rapid detection of acute kidney injury by plasma and urinary neutrophil gelantinase-associated lipocalin after cardiopulmonary bypass. J Cardiovasc Pharmacol 2009;53:261–266.

13 Bennett M, Dent CL, Ma Q, Dastrala S, Grenier F, Workman R, Syed H, Ali S, Barasch J, Devarajan P: Urine NGAL predicts severity of acute kidney injury after cardiac surgery: a prospective study. Clin J Am Soc Nephrol 2008;3:665–673.

14 Thakar CV, Yared JP, Worley S, Cotman K, Paganini EP: Renal dysfunction and serious infection after open-heart surgery. Kidney Int 2003;64:239–246.

15 Jiang M, Dong Z: Regulation and pathological role of p53 in cisplatin nephrotoxicity. J Pharmacol Exp Ther 2008;327:300–307.

16 National Heart, Lung, and Blood Institute Acute Respiratory Distress Syndrome (ARDS) Clinical Trials Network, Wiedemann HP, Wheeler AP, Bernard GR, Thompson BT, Hatden D, deBoisblanc B, Connors AF Jr, Hite RD, Harabin AL: Comparison of two fluid-management strategies in acute lung injury. N Engl J Med 2006;354:2564–2575.

17 Btaiche IF, Mohammad RA, Alaniz C, Mueller BA: Amino acid requirements in critically ill patients with acute kidney injury treated with continuous renal replacement therapy. Pharmacotherapy 2008;28:600–613.

18 Cruz DN, Antonelli M, Fumagalli R, Brienza N, Donati A, Malcangi V, Petrini F, Volta G, Bobbio Pallavicini FM, Rottoli F, Giunta F, Ronco C: Early use of polymyxin B hemoperfusion in abdominal septic shock: the EUPHAS randomized controlled trial. JAMA 2009;301:2445–2452.

19 Antonelli M, Fumagalli R, Cruz DN, Brienza N, Giunta F, EUPHAS Study Group: PMX Endotoxin Removal in the clinical practice: results from the EUPHAS Trial; in Ronco C, Piccinni P, Rosner MH: Endotoxemia and Endotoxin Shock: Disease, Diagnosis and Therapy. Contrib Nephrol. Basel, Karger, 2010, vol 167, pp 83–90.

20 Ronco C, Brendolan A, Dan M, Piccinni P, Bellomo R, DeNitti C, Inguaggiato P, Tetta C: Adsorption in sepsis. Kidney Int Suppl 2000;76:S148–S155.

21 Peng Z, Singbartl K, Simon P, Rimmele T, Bishop J, Clermont G, Kellum JA: Blood purification in sepsis: a new paradigm; in Ronco C, Bellomo R, McCullough PA (eds): Cardiorenal Syndromes in Critical Care. Contrib Nephrol. Basel, Karger, 2010, vol 165, pp 322–328.

22 Kimmel JD, Gibson GA, Watkins SC, Kellum JA, Federspiel WJ: IL-6 adsorption dynamics in hemoadsorption beads studied using confocal laser scanning microscopy. J Biomed Mater Res B Appl Biomater 2010;92:390–396.

23 Ronco C, Piccinni P, Kellum J: Rationale of extracorporeal removal of endotoxin in sepsis: theory, timing and technique; in Ronco C, Piccinni P, Rosner MH (eds): Endotoxemia and Endotoxin Shock: Disease, Diagnosis and Therapy. Contrib Nephrol. Basel, Karger, 2010, vol 167, pp 25–34.

24 Bellomo R, Tetta C, Ronco C: Coupled plasma filtration adsorption. Intensive Care Med 2003;29:1222–1228.

25 Humes HD, Weitzel WF, Bartlett RH, Swaniker FC, Paganini EP, Luderer JR, Sobota J: Initial clinical results of the bioartificial kidney containing human cells in ICU patients with acute renal failure. Kidney Int 2004;66:1578–1588.

26 Ding F, Humes HD: The bioartificial kidney and bioengineered membranes in acute kidney injury. Nephron Exp Nephrol 2008;109:118–122.

27 Muthusubramaniam L, Lowe R, Fissell WH, Li L, Marchant RE, Desai TA, Roy S: Hemocompatibility of silicon-based substrates for biomedical implant applications. Ann Biomed Eng 2011;39:1296–1305.

28 Li L, Marchant RE, Dubnisheva A, Roy S, Fissell WH: Anti-biofouling sulfobetaine polymer thin films on silicon and silicon nanopore membranes. J Biomater Sci Polym Ed 2010, E-pub ahead of print.

29 Srisawat N, Lawsin L, Uchino S, Bellomo R, Kellum JA, BEST Kidney Investigators: Cost of acute renal replacement therapy in the intensive care unit: results from the Beginning and Ending Supportive Therapy for the Kidney (BEST KIDNEY) study. Crit Care 2010;14:R46.

Emil P. Paganini, MD, FACP, FRCP
Propags Consulting, LLC
8700 Mayfield Road, Bld 'A'
Chesterland, Ohio 44026 USA
Tel: +1 440 729 2452, E-Mail epaganini@propags.com

Author Index